Neverending Fracti

T0297302

Despite their classical nature, continued fractions are a neverending research area, with a body of results accessible enough to suit a wide audience, from researchers to students and even amateur enthusiasts. *Neverending Fractions* brings these results together, offering fresh perspectives on a mature subject.

Beginning with a standard introduction to continued fractions, the book covers a diverse range of topics, including elementary and metric properties, quadratic irrationals and more exotic topics such as folded continued fractions and Somos sequences. Along the way, the authors reveal some amazing applications of the theory to seemingly unrelated problems in number theory. Previously scattered throughout the literature, these applications are brought together in this volume for the first time.

A wide variety of exercises guide readers through the material, and this will be especially helpful to readers using the book for self-study; the authors also provide many pointers to the literature.

Australian Mathematical Society Lecture Series: 23

Neverending Fractions
An Introduction to Continued Fractions

JONATHAN BORWEIN
University of Newcastle, New South Wales

ALF VAN DER POORTEN
Macquarie University, Sydney

JEFFREY SHALLIT
University of Waterloo, Ontario

WADIM ZUDILIN
University of Newcastle, New South Wales

CAMBRIDGE
UNIVERSITY PRESS

CAMBRIDGE
UNIVERSITY PRESS

University Printing House, Cambridge CB2 8BS, United Kingdom

Cambridge University Press is part of the University of Cambridge.

It furthers the University's mission by disseminating knowledge in the pursuit of
education, learning and research at the highest international levels of excellence.

www.cambridge.org
Information on this title: www.cambridge.org/9780521186490

© Jonathan Borwein, The Estate of Alf van der Poorten, Jeffrey Shallit
and Wadim Zudilin 2014

First published 2014

A catalogue record for this publication is available from the British Library

Library of Congress Cataloguing in Publication data
Borwein, Jonathan M.
Neverending fractions: an introduction to continued fractions / Jonathan Borwein,
University of Newcastle, New South Wales, Alf van der Poorten, Macquarie University,
Sydney, Jeffrey Shallit, University of Waterloo, Ontario, Wadim Zudilin, University of
Newcastle, New South Wales.
pages cm – (Australian Mathematical Society lecture series; 23)
Includes bibliographical references and index.
ISBN 978-0-521-18649-0 (alk. paper)
1. Continued fractions. 2. Processes, Infinite. 3. Fractions.
I. Van Der Poorten, A. J. II. Shallit, Jeffrey Outlaw. III. Zudilin, Wadim.
IV. Title. V. Title: Continued fractions.
QA295.B667 2014
513.2′6–dc23
2014006908

ISBN 978-0-521-18649-0 Paperback

Additional resources for this publication at http://carma.newcastle.edu.au/Neverending

Contents

Preface

This book arose from many lectures the authors delivered independently at different locations to students of different levels.

'Theory' is a scientific name for 'story'. So, if the reader somehow feels uncomfortable about following a theory of continued fractions, he or she might be more content to read the story of *neverending* fractions.

The queen of mathematics – number theory – remains one of the most accessible parts of significant mathematical knowledge. Continued fractions form a classical area within number theory, and there are many textbooks and monographs devoted to them. Despite their classical nature, continued fractions remain a neverending research field, many of whose results are elementary enough to be explained to a wide audience of graduates, postgraduates and researchers, as well as teachers and even amateurs in mathematics. These are the people to whom this book is addressed.

After a standard introduction to continued fractions in the first three chapters, including generalisations such as continued fractions in function fields and irregular continued fractions, there are six 'topics' chapters. In these we give various amazing applications of the theory (irrationality proofs, generating series, combinatorics on words, Somos sequences, Diophantine equations and many other applications) to seemingly unrelated problems in number theory. The main feature that we would like to make apparent through this book is the naturalness of continued fractions and of their expected appearance in mathematics. The book is a combination of formal and informal styles. The aforementioned applications of continued fractions are, for the most part, not to be found in earlier books but only in scattered scientific articles and lectures.

We have included various remarks and exercises but have been sparing with the latter. In the topics chapters we do not always give full details. Needless to say, all topics can be followed up in the end notes for each chapter and through the references.

We would like to thank many colleagues for useful conversations during the development of this book, especially Mumtaz Hussain, Pieter Moree and James Wan. We are also deeply indebted to the copy-editor Susan Parkinson whose incisive and tireless work on the book has enhanced its appearance immeasurably.

Finally, Alf van der Poorten (1942–2010) died before this book could be brought to fruition. He was both a good friend and a fine colleague. We offer this book both in his memory and as a way of bringing to a more general audience some of his wonderful contributions to the area. Chapters 4, 5 and 6 originate in lectures Alf gave in the last few years of his life and, for matters of both taste and necessity, they are largely left as presentations in his unique and erudite style.

Alfred "Alf" Jacobus van der Poorten
(16 May 1942 in Amsterdam–9 October 2010 in Sydney)

A full mathematical biography of Alf is to be found in the 2013 volume dedicated to his memory [31].

Jon Borwein, Jeffrey Shallit, and Wadim Zudilin
Newcastle and Waterloo

1

Some preliminaries from number theory

In this chapter we provide the necessary prerequisites from multiplicative number theory regarding primes, divisibility and approximation by rationals.

1.1 Divisibility in $\mathbb{Z}$. Euclidean algorithm

The basic objects of our story are the set of natural numbers $\mathbb{N} = \{1, 2, 3, \ldots\}$ and the set of integers $\mathbb{Z}$. In addition, we often deal with the set of *rationals* $\mathbb{Q}$ and the set of real numbers $\mathbb{R}$. An element of $\mathbb{R} \setminus \mathbb{Q}$ is called *irrational*. Shortly we will need the complex numbers $\mathbb{C}$ as well.

The set of integers $\mathbb{Z}$ forms a *ring* equipped with the usual addition and multiplication. The operation of division, the inverse to multiplication, applies to pairs (a, b) with $b \neq 0$. We say that a number $b \neq 0$ *divides* a (writing $b \mid a$) or, equivalently, b is a *divisor* of a or a is *divisible* by b or a is a *multiple* of b, if $a = bq$ holds for some integer q. The number q is called the *quotient* of a by b. The number 0 is divisible by any integer $b \neq 0$. If $a \neq 0$ then the number of its divisors is finite. We use the notation $b \nmid a$ to say that b does not divide a.

Let us list some simple properties of divisibility in $\mathbb{Z}$.

Lemma 1.1 *If $c \mid b$ and $b \mid a$ then $c \mid a$.*

Proof Since $b = cq_1$ and $a = bq_2$, we have $a = c(q_1 q_2)$. □

Lemma 1.2 *If all terms in an equality $a_1 + \cdots + a_n = b_1 + \cdots + b_k$, except one, are multiples of a fixed integer c then the exceptional term is a multiple of c as well.*

Proof Writing all terms except b_k, say, in the form $a_i = c\widetilde{a}_i$ and $b_j = c\widetilde{b}_j$, we see that

$$b_k = c(\widetilde{a}_1 + \cdots + \widetilde{a}_n - \widetilde{b}_1 - \cdots - \widetilde{b}_{k-1}).$$

1

This means that b_k can be represented in the form $b_k = cb$ for some integer b, and hence b_k is a multiple of c. □

The *floor* or *greatest integer function* is denoted $\lfloor x \rfloor$ and is defined to be the greatest integer $\le x$. Thus $\lfloor \pi \rfloor = 3$ and $\lfloor -e \rfloor = -3$. The *ceiling* or *least integer function* $\lceil x \rceil$ is defined, analogously, to be the least integer $\ge x$. Clearly $\lfloor x \rfloor \le x < \lfloor x \rfloor + 1$. We let $\{x\}$ denote the *fractional part* of x, that is, $x - \lfloor x \rfloor$; hopefully there will be no confusion with ordinary set notation.

Theorem 1.3 (Division with remainder) *For any integer a and any positive integer b, there exist integers q and r such that*

$$a = bq + r \quad and \quad 0 \le r < b. \tag{1.1}$$

These numbers q and r are defined uniquely.

Proof The existence of such a pair q, r is clear; we take $q = \lfloor a/b \rfloor$. Then $q \le a/b < q + 1$; hence $bq \le a < b(q + 1)$ and so $0 \le r = a - bq < b$.

Assuming two representations (1.1), the second being of the form

$$a = bq_1 + r_1 \quad and \quad 0 \le r_1 < b, \tag{1.2}$$

we deduce from equations (1.1) and (1.2) that

$$0 = b(q - q_1) + (r - r_1) \quad and \quad |r - r_1| < b. \tag{1.3}$$

Hence b divides the difference $r - r_1$, which is possible only when $r = r_1$, by the inequality in (1.3). The equality $r = r_1$ implies $q = q_1$ by the equality in (1.3). □

An integer dividing each of the integers $a_1, a_2, \ldots, a_n$ is called their *common divisor*; the largest of the common divisors is called the *greatest common divisor* and denoted by $\gcd(a_1, a_2, \ldots, a_n)$. If $\gcd(a_1, \ldots, a_n) = 1$, the numbers $a_1, \ldots, a_n$ are called *coprime* (or *relatively prime*). If every pair of the set $\{a_1, \ldots, a_n\}$ is coprime then the set is called *pairwise coprime*. (The latter requirement is stronger, as the example of the set $\{6, 10, 15\}$ shows: these numbers are coprime but not pairwise coprime.)

The following two lemmas can be easily verified using the above definitions.

Lemma 1.4 *If a is a multiple of b then the set of common divisors of a and b coincides with the set of divisors of b; in particular, $\gcd(a, b) = |b|$.*

Lemma 1.5 *If $a = bq + r$ then the set of common divisors of a and b coincides with the set of common divisors of b and r; in particular, $\gcd(a, b) = \gcd(b, r)$.*

The last statement substantiates the following classical Greek algorithm for computing the greatest common divisor of two numbers $a, b \in \mathbb{Z}$:

EUCLIDEAN ALGORITHM   Let $a, b \in \mathbb{Z}$ with $a \geq b > 0$. Defining $r_{-1} = a$ and $r_0 = b$, consider the following successive application of division with remainder (Theorem 1.3):

$$
\begin{aligned}
r_{-1} &= r_0 q_0 + r_1, & 0 < r_1 &< r_0, \\
r_0 &= r_1 q_1 + r_2, & 0 < r_2 &< r_1, \\
&\ \ \vdots & & \\
r_{n-1} &= r_n q_n + r_{n+1}, & 0 < r_{n+1} &< r_n, \\
r_n &= r_{n+1} q_{n+1}.
\end{aligned}
\tag{1.4}
$$

Then the last nonzero remainder r_{n+1} is the greatest common divisor of a and b.

Critically, the procedure (1.4) terminates at some step in view of the following chain of inequalities:

$$r_0 > r_1 > \cdots > r_{n-1} > r_n > r_{n+1} > 0.$$

By (1.4) and Lemma 1.5 we get

$$
\begin{aligned}
\gcd(a, b) &= \gcd(r_{-1}, r_0) = \gcd(r_0, r_1) = \gcd(r_1, r_2) = \cdots \\
&= \gcd(r_n, r_{n+1}) = \gcd(r_{n+1}, 0) = r_{n+1}.
\end{aligned}
$$

Hence the last nonzero remainder r_{n+1} in (1.4) is indeed the required greatest common divisor of a and b.

Lemma 1.6   *For any integer* $m > 0$ *we have* $\gcd(am, bm) = m \gcd(a, b)$.

Proof   Multiply all equalities in (1.4) by m.   □

Lemma 1.7   *Let* δ *be a common divisor of* a *and* b. *Then*

$$\gcd\left(\frac{a}{\delta}, \frac{b}{\delta}\right) = \frac{\gcd(a, b)}{|\delta|}$$

and, in particular,

$$\gcd\left(\frac{a}{\gcd(a, b)}, \frac{b}{\gcd(a, b)}\right) = 1.$$

Proof   By Lemma 1.6 we obtain

$$\gcd(a, b) = \gcd\left(\frac{a}{\delta}\delta, \frac{b}{\delta}\delta\right) = \gcd\left(\frac{a}{\delta}, \frac{b}{\delta}\right)|\delta|.   □$$

We leave the proofs of Lemmas 1.8–1.10 below to the reader.

Lemma 1.8   *If* $\gcd(a, b) = 1$ *then* $\gcd(ac, b) = \gcd(c, b)$.

4 Some preliminaries from number theory

Lemma 1.9 *If* $\gcd(a, b) = 1$ *and* ac *is divisible by* b *then* c *is divisible by* b.

Lemma 1.10 *If each of the numbers* $a_1, \ldots, a_n$ *is coprime with each of the numbers* $b_1, \ldots, b_k$ *then the products* $a_1 \cdots a_n$ *and* $b_1 \cdots b_k$ *are coprime.*

An integer that is a multiple of all the numbers $a_1, \ldots, a_n$ is called their *common multiple*. The smallest positive common multiple is called the *least common multiple* or *lcm* and denoted by $\mathrm{lcm}(a_1, \ldots, a_n)$.

Lemma 1.11 *The set of common multiples of two given numbers coincides with the set of multiples of their least common multiple.*

Proof Let M denote a common multiple of the given integers a and b. Then $M = ak$ for $k \in \mathbb{Z}$, since M is a multiple of a, and the number $M/b = ak/b$ is an integer. Define $d = \gcd(a, b)$, $a = da_1$ and $b = db_1$; by Lemma 1.7 we have $\gcd(a_1, b_1) = 1$. By Lemma 1.9 the equality $M/b = a_1 k/b_1 \in \mathbb{Z}$ implies that k is divisible by b_1, that is, $k = b_1 t = bt/d$ for $t \in \mathbb{Z}$. Therefore

$$M = \frac{ab}{d}t = \frac{ab}{\gcd(a, b)}t, \qquad t \in \mathbb{Z}, \tag{1.5}$$

and, as can be seen, any such M is a multiple of both a and b. We get the least common multiple by specialization $t = \pm 1$: $\mathrm{lcm}(a, b) = |ab|/\gcd(a, b)$. Thus, formula (1.5) can be written in the required form $M = \mathrm{lcm}(a, b)t$ with $t \in \mathbb{Z}$. □

The previous lemma gives a simple and efficient algorithm for computing the least common multiple for a set $a_1, a_2, \ldots, a_n$ of arbitrary length $n \geq 2$. Namely, we have the formula

$$\mathrm{lcm}(a_1, a_2, a_3, a_4, \ldots, a_n) = \mathrm{lcm}(\mathrm{lcm}(\ldots \mathrm{lcm}(\mathrm{lcm}(\mathrm{lcm}(a_1, a_2), a_3), a_4), \ldots), a_n),$$

while the least common multiple of just two numbers is computed by

$$\mathrm{lcm}(a, b) = \frac{|ab|}{\gcd(a, b)}.$$

EXERCISE 1.12 Show that, for a pair of relatively prime integers a and b, the linear equation $ax - by = 1$ has infinitely many solutions in integers x, y.

Hint This can be split into two parts: First, show (using either an inductive argument or the Euclidean algorithm) that there exists at least one solution of the equation, say x_0, y_0, and, second, that the pair $x = x_0 + bt$, $y = y_0 + at$ is a solution for any $t \in \mathbb{Z}$. □

1.2 Primes

An integer exceeding 1 always has at least two distinct divisors, namely, 1 and itself. If these two divisors exhaust the list of all positive divisors of such an integer then the integer is called a *prime number*; otherwise, the integer (> 1) is called a *composite* number.

Lemma 1.13 *The least positive divisor, different from* 1, *of an integer a* > 1 *is a prime.*

Proof The set $A = \{2, 3, \ldots, a\}$ is not empty and finite and contains at least one divisor (namely, a) of the given integer a; thus we can choose the smallest such divisor, say b. If b is not prime then it has a divisor c such that $1 < c < b$, so that $c \in A$. But then Lemma 1.1 implies that c divides a, which contradicts our choice of b. □

The next lemma, while simple, is very potent.

Lemma 1.14 *The least positive divisor, different from* 1, *of a* composite *integer a* > 1 *does not exceed* $\sqrt{a}$.

Proof Let $b > 1$ be the least positive divisor of a. Write $a = bc$; since a is composite we have $b < a$, so that $c > 1$. As both b and c are divisors of a and b is the least divisor we have $b \le c = a/b$, implying that $b^2 \le a$. □

The next result, attributed to Euclid of Alexandria, circa 300 BCE, illustrates the sophistication of Greek number theory.

Theorem 1.15 (Euclid) *The set of primes is infinite.*

Proof If not, we could write the (nonempty) set of primes as

$$\{p_1 = 2, p_2 = 3, p_3, \ldots, p_n\}$$

and consider the least positive divisor, different from 1, of the number

$$p_1 p_2 \cdots p_n + 1.$$

The divisor is prime, by Lemma 1.13, and it is not on our list because it is relatively prime to each of $p_1, \ldots, p_n$. Thus, we arrive at a contradiction. □

An important property (as well as the main difficulty in use) of primes is their role as 'building blocks' or 'atoms' in the study of $\mathbb{Z}$ from the multiplicative point of view.

Lemma 1.16 *Every integer a is either a multiple of a given prime p or co-prime with p.*

Proof   Indeed, $\gcd(a, p)$ is p or 1, as a divisor of p.   □

Lemma 1.17   *If a product of some terms is divisible by p then at least one of the terms is divisible by p.*

Proof   Otherwise, each term is coprime with p by Lemma 1.16, while Lemma 1.10 implies that the product has to be coprime with p as well.   □

Theorem 1.18 (Fundamental theorem of arithmetic)   *Every integer greater than 1 may be decomposed into a product of primes (that is, factorised), and this decomposition is unique (up to the ordering of the primes in it).*

Proof   *Existence*. This is shown by induction on $a > 1$. For the number $a = 2$, its factorisation is trivial (owing to the primality of 2). If $a > 2$ then it is either a prime (and hence its factorisation involves only the number itself) or composite. In the latter case it can be written in the form $a = pa_1$, where p is the prime divisor from Lemma 1.13, and for the number a_1, $\sqrt{a} \le a_1 < a$, we use the induction hypothesis.

Uniqueness. Assume, contrary to what we want to prove, that numbers with non-unique factorisation exist, and choose the least in the set of such numbers, say a:

$$a = p_1 p_2 \cdots p_n = q_1 q_2 \cdots q_k, \tag{1.6}$$

$p_1, p_2, \ldots, p_n$ and $q_1, q_2, \ldots, q_k$ are primes.

On the one hand, the right-hand side of (1.6) is divisible by q_1, and hence at least one term on the left-hand side of (1.6) (say p_1, without loss of generality) is divisible by q_1 by Lemma 1.17. On the other hand, each term on the left-hand side of (1.6) is a prime and therefore p_1 has to coincide with q_1; after reduction by $p_1 = q_1$ in (1.6) we obtain

$$p_2 \cdots p_n = q_2 \cdots q_k. \tag{1.7}$$

At least one side of (1.7) involves a non-empty product (otherwise we would have $a = p_1 = q_1$, two identical prime factorisations of the number a, contradicting its choice above). Thus, (1.7) records two different factorisations of a number a_1 satisfying $1 < a_1 < a$. The latter contradicts the minimality of our choice of a.   □

Two very important problems in number theory, with numerous applications to the theory and practice of information security and encryption, are deciding whether a given number is prime and finding large prime numbers. The latter problem is related to the distribution of primes in the set of positive integers. In fact, it is not hard to show that there are arbitrarily long sequences of consecutive composite numbers (for example, the sequence $n! + i$ for $i = 2, 3, \ldots, n$). However, it is conjectured that there are infinitely many *twin primes*, that is, infinitely many pairs p, q of primes with $q - p = 2$. Very recent (2013) work by Zhang [173], as yet unpublished, has proved that there are infinitely many pairs of primes that differ by some $N < 70\,000\,000$; subsequently his methods were refined by the Polymath project to $N \leq 4680$ and by Maynard to $N \leq 600$. The latter number may seem feeble as a replacement for 2 but in fact it is an enormous accomplishment.

Another famous conjecture ('Goldbach's conjecture') states that any even integer greater than 2 is a sum of two primes. The recent work of Helfgott [73] reports on a proof of its weaker three-primes version.

The following result, known as the *prime number theorem*, is a fundamental theorem on the distribution of primes. It was almost guessed, from much numerical evidence, by Gauss in 1791; Chebyshev provided some evidence for it in 1850, and finally Hadamard and de la Vallée Poussin independently proved it in 1896 using methods of complex analysis. Chebyshev's work was good enough to prove *Bertrand's postulate*: there is a prime in the interval $[n - 1, 2n - 1]$ for each $n \geq 3$.

The main feature of the proofs of Hadamard and de la Vallée Poussin is the use of the Riemann zeta function $\zeta(s)$, defined in the complex half-plane $\operatorname{Re} s > 1$ by the (slowly convergent for small $s > 1$) series

$$\zeta(s) = \sum_{n=1}^{\infty} \frac{1}{n^s}. \tag{1.8}$$

Theorem 1.19 (Prime number theorem) *Let $\pi(x)$ be the number of primes less than or equal to x, for any real number x. Then*

$$\lim_{x \to \infty} \frac{\pi(x)}{x/\log x} = 1,$$

where $\log x = \ln x$ denotes the logarithm of x to the base e.

The details of this proof are a little tangential to our goals (the interested reader may find them in many places including [77, Chapter II]), but the following curious equivalent form of the prime number theorem will be useful later.

Theorem 1.20 (Rate of growth of lcm) *Let $d_n = \operatorname{lcm}(1, 2, \ldots, n)$ be the least common multiple of the first n consecutive natural numbers. Then*

$$\lim_{n \to \infty} \frac{\log d_n}{n} = 1.$$

It is clear that $n!$ is a common multiple of the numbers $1, 2, \ldots, n$, and it grows as $(n/e)^n \sqrt{2\pi n}(1 + o(1))$ according to *Stirling's asymptotic formula* (which can be replaced by rougher estimates that we will establish below in (2.37)). Theorem 1.20 tells us that the actual growth of the *least* common multiple in this case is, roughly speaking, e^n, which is of course asymptotically a better estimate than the one arising from $n!$.

EXERCISE 1.21 (see, for example, [77, Chapter I, Theorem 3]) Show the equivalence of Theorems 1.19 and 1.20.

1.3 Fibonacci numbers and the complexity of the Euclidean algorithm

The sequence of *Fibonacci numbers* $F_0, F_1, F_2, F_3, \ldots$ is defined by the simple linear recurrence relation

$$F_{n+2} = F_{n+1} + F_n \tag{1.9}$$

and the initial data $F_0 = 0$, $F_1 = 1$. It is a sequence on which the Euclidean algorithm (see the text after Lemma 1.5)

$$
\begin{aligned}
a &= bq_0 + r_1, & 0 &< r_1 < b, \\
b &= r_1 q_1 + r_2, & 0 &< r_2 < r_1, \\
r_1 &= r_2 q_2 + r_3, & 0 &< r_3 < r_2, \\
&\;\;\vdots \\
r_{n-1} &= r_n q_n + r_{n+1}, & 0 &< r_{n+1} < r_n, \\
r_n &= r_{n+1} q_{n+1}
\end{aligned}
\tag{1.10}
$$

(here $a \geq b > 0$) works for more steps than might be expected: since the quotients $q_0, q_1, \ldots, q_{n+1}$ are integers greater than or equal to 1, the following estimates hold:

$$r_{n+1} \geq F_1, \quad r_n \geq F_2, \quad r_{n-1} \geq F_3, \quad \ldots,$$

$$r_1 \geq F_{n+1}, \quad b \geq F_{n+2}, \quad a \geq F_{n+3}.$$

Lemma 1.22 *If k is the number of steps (divisions with remainder) in the Euclidean algorithm then for given initial data $a \geq b > 0$ we have $a \geq F_{k+1}$ and $b \geq F_k$.*

Our immediate aim is to deduce a general form for the Fibonacci sequence.

This will allow us to give an upper bound on the number of steps in the Euclidean algorithm (the *complexity* of the algorithm) for arbitrary initial data $a \geq b > 0$.

EXERCISE 1.23   Before moving to the rest of this section, find and prove a closed-form expression for the Fibonacci numbers.

Now let $a_1(n), a_2(n), \ldots, a_m(n)$ be arbitrary functions of the nonnegative integer argument n. The recurrence equation

$$\phi(n + m) + a_1(n)\phi(n + m - 1) + \cdots + a_{m-1}(n)\phi(n + 1) + a_m(n)\phi(n) = 0 \quad (1.11)$$

is called a *linear homogeneous difference equation of order m*, and any function $\phi(n)$ satisfying (1.11) for all $n = 0, 1, 2, \ldots$ is called its *solution*. It is not difficult to see that the choice of *initial data*

$$\phi(0) = \phi_0, \quad \phi(1) = \phi_1, \quad \ldots, \quad \phi(m - 1) = \phi_{m-1}$$

determines a solution $\phi(n)$, $n = 0, 1, 2, \ldots$, of (1.11) uniquely.

Lemma 1.24   *Let* $\phi^{(1)}(n), \phi^{(2)}(n), \ldots, \phi^{(k)}(n)$ *be k solutions of* (1.11). *Then the function*

$$\phi(n) = c_1\phi^{(1)}(n) + c_2\phi^{(2)}(n) + \cdots + c_k\phi^{(k)}(n), \qquad n = 0, 1, 2, \ldots,$$

where $c_1, c_2, \ldots, c_k$ *are arbitrary constants from the ground field (for example,* $\mathbb{Q}$ *or* $\mathbb{R}$), *is a solution of* (1.11) *as well.*

Equivalently, the set of solutions of (1.11) forms a linear space. Furthermore, we can always construct m linearly independent solutions of the equation, $\phi^{(1)}(n), \phi^{(2)}(n), \ldots, \phi^{(m)}(n)$, by choosing the initial data in such a way that the m-vectors

$$\begin{pmatrix} \phi_0^{(1)} \\ \phi_1^{(1)} \\ \vdots \\ \phi_{m-1}^{(1)} \end{pmatrix}, \quad \begin{pmatrix} \phi_0^{(2)} \\ \phi_1^{(2)} \\ \vdots \\ \phi_{m-1}^{(2)} \end{pmatrix}, \quad \ldots, \quad \begin{pmatrix} \phi_0^{(m)} \\ \phi_1^{(m)} \\ \vdots \\ \phi_{m-1}^{(m)} \end{pmatrix}$$

are linearly independent.

Lemma 1.25   *A general solution of* (1.11) *can be written in the form*

$$\phi(n) = c_1\phi^{(1)}(n) + c_2\phi^{(2)}(n) + \cdots + c_m\phi^{(m)}(n), \qquad n = 0, 1, 2, \ldots,$$

where $\phi^{(1)}(n), \phi^{(2)}(n), \ldots, \phi^{(m)}(n)$ *is a fixed basis (defined above) in the solution space, while* $c_1, \ldots, c_m$ *are arbitrary constants.*

Proof Let $\phi(n)$, $n = 0, 1, 2, \ldots$, be a solution of (1.11). Then the constants $c_1, \ldots, c_m$ are determined by the system of linear equations

$$c_1 \begin{pmatrix} \phi_0^{(1)} \\ \phi_1^{(1)} \\ \vdots \\ \phi_{m-1}^{(1)} \end{pmatrix} + c_2 \begin{pmatrix} \phi_0^{(2)} \\ \phi_1^{(2)} \\ \vdots \\ \phi_{m-1}^{(2)} \end{pmatrix} + \cdots + c_m \begin{pmatrix} \phi_0^{(m)} \\ \phi_1^{(m)} \\ \vdots \\ \phi_{m-1}^{(m)} \end{pmatrix} = \begin{pmatrix} \phi(0) \\ \phi(1) \\ \vdots \\ \phi(m-1) \end{pmatrix}. \qquad \square$$

EXERCISE 1.26 Prove Lemma 1.24 and finalise the proof of Lemma 1.25.

From now on we switch to the simplest case, when the coefficients $a_1, \ldots, a_{m-1}, a_m$ of the difference equation do not depend on n and, in addition, the *characteristic polynomial*

$$\lambda^m + a_1 \lambda^{m-1} + \cdots + a_{m-1} \lambda + a_m = 0 \qquad (1.12)$$

of (1.11) has exactly m distinct *nonzero* roots $\lambda_1, \ldots, \lambda_m$. (For the case of repeated roots, we recommend [20].)

Theorem 1.27 (Solution to recursion with no repeated roots) *A general solution of the linear homogeneous difference equation with constant coefficients has the form*

$$\phi(n) = c_1 \lambda_1^n + \cdots + c_m \lambda_m^n, \qquad n = 0, 1, 2, \ldots.$$

Proof Note that the functions $\phi^{(j)}(n) = \lambda_j^n$, where $j = 1, \ldots, m$, form a *fundamental solution system*, that is, a basis in the solution space. The fact that the solutions are linearly independent follows from the nonvanishing of a Vandermonde determinant (see, for example, [89, Section 2.1] for more information about the latter). $\qquad \square$

Lemma 1.28 *The Fibonacci numbers are also given by the explicit formula*

$$F_n = \frac{\alpha^n - \beta^n}{\sqrt{5}},$$

where $\alpha = (1 + \sqrt{5})/2$ and $\beta = (1 - \sqrt{5})/2$ (with $\alpha\beta = -1$).

Proof Indeed, the characteristic polynomial $\lambda^2 - \lambda - 1$ of the difference equation (1.9) has roots α, β. Letting

$$F_n = c_1 \alpha^n + c_2 \beta^n$$

and setting $n = 0$ and $n = 1$, we find $c_1 = -c_2 = 1/\sqrt{5}$. $\qquad \square$

EXERCISE 1.29 (Pell numbers) The Pell numbers satisfy the recurrence relation $P_{n+2} = 2P_{n+1} + P_n$ with initial conditions $P_0 = 0$ and $P_1 = 1$. Give a closed-form expression for the Pell numbers.

Theorem 1.30 (Complexity of Euclidean algorithm) *The number k of steps in the Euclidean algorithm with data a ≥ b > 0 is bounded as follows:*

$$k < 2.5 \log a + 1.5.$$

Proof By Lemmas 1.22 and 1.28 we have

$$a \geq F_{k+1} \geq \frac{1}{\sqrt{5}}\left(\left(\frac{1+\sqrt{5}}{2}\right)^{k+1} - 1\right),$$

implying that

$$\left(\frac{1+\sqrt{5}}{2}\right)^{k+1} \leq a\sqrt{5} + 1$$

and

$$k \leq \frac{\log(a\sqrt{5}+1)}{\log((1+\sqrt{5})/2)} - 1 \leq \frac{\log(a(\sqrt{5}+1))}{\log((1+\sqrt{5})/2)} - 1$$
$$< \frac{1.18 + \log a}{0.48} - 1 < 2.5 \log a + 1.5. \qquad \square$$

1.4 Approximation of real numbers by rationals

Let α be a real number. Studying the behaviour of the difference

$$\left|\alpha - \frac{p}{q}\right| \qquad (1.13)$$

as a function of $p \in \mathbb{Z}$ and $q \in \mathbb{N}$ plays an important role in number theory and its applications. Since $\mathbb{Q}$ is everywhere dense in $\mathbb{R}$, we can always choose p and q such that the quantity (1.13) is as small as required. That is why it is interesting to study the *relative* smallness of (1.13), namely, how small this quantity could become as the parameters p and q, with $q \leq q_0$, vary.

Let $\phi(q)$ be a positive function of $q \in \mathbb{N}$ that tends to 0 as q tends to infinity. We say that a number $\alpha \in \mathbb{R}$ is *approximable to order $\phi(q)$ by rational numbers* p/q if there exists a constant $c > 0$ depending on α and $\phi(q)$ such that the inequality

$$\left|\alpha - \frac{p}{q}\right| < c\phi(q) \qquad (1.14)$$

has infinitely many solutions in $p \in \mathbb{Z}$ and $q \in \mathbb{N}$ with $p/q \neq \alpha$. We say that $\phi(q)$ is *the best approximation order of α by rationals* if α is approximable to order $\phi(q)$ and, for a certain constant $c_1 > 0$ depending on α and $\phi(q)$, the inequality

$$\left| \alpha - \frac{p}{q} \right| < c_1 \phi(q) \tag{1.15}$$

has at most finitely many solutions in $p \in \mathbb{Z}$ and $q \in \mathbb{N}$ with $p/q \neq \alpha$.

EXERCISE 1.31   Show that the latter condition, (1.15), implies the existence of a constant $c_2 > 0$ with the property that

$$\left| \alpha - \frac{p}{q} \right| > c_2 \phi(q)$$

holds for all $p \in \mathbb{Z}$ and $q \in \mathbb{N}$ with $p/q \neq \alpha$.

A standard choice of the function $\phi(q)$ is the power function $\phi(q) = q^{-\nu}$, where $\nu > 0$. Letting ν and c take different values, a central problem in this case is to determine whether (1.14) has finitely or infinitely many solutions.

With this special choice of $\phi(q) = q^{-\nu}$, we say in short that α is *approximable to order* ν and that ν is *the best approximation order of α by rationals*, respectively. The infimum over real $\nu > 0$ such that the inequality

$$0 < \left| \alpha - \frac{p}{q} \right| \leq \frac{1}{q^\nu}$$

has at most finitely many solutions in $p \in \mathbb{Z}$ and $q \in \mathbb{N}$ is commonly called the *irrationality exponent* or *irrationality measure* of α and is denoted by $\mu(\alpha)$. If a number has infinite irrationality measure, it is called a *Liouville number*.

Lemma 1.32   *If $\alpha \in \mathbb{Q}$, that is, $\alpha = a/b$ with $\gcd(a,b) = 1$, then the inequality*

$$\left| \alpha - \frac{p}{q} \right| < \frac{c}{q} \tag{1.16}$$

has no solutions in $p/q \neq \alpha$ for any c satisfying $0 < c \leq 1/b$.

Proof   Indeed,

$$\left| \alpha - \frac{p}{q} \right| = \left| \frac{a}{b} - \frac{p}{q} \right| = \frac{|aq - bp|}{bq} \geq \frac{1}{bq}, \tag{1.17}$$

and the lemma follows.   □

Lemma 1.33   *If $\alpha \in \mathbb{Q}$, that is, $\alpha = a/b$ with $\gcd(a,b) = 1$, then the inequality* (1.16) *has infinitely many solutions $p/q \neq \alpha$ for any c satisfying $c > 1/b$.*

Proof   With the help of Exercise 1.12 above we see that there are infinitely many integers p and q (even when we restrict q to be at least 1) of the equation $ap - bq = 1$. The numbers p and q in any such solution are coprime and satisfy $p/q \neq \alpha$ and the inequality (1.16) for any $c > 1/b$.   □

Theorem 1.34 (Irrationality criterion) *For a given $\alpha \in \mathbb{R}$, assume that there exist $\epsilon > 0$ and a sequence of rational fractions $(p_n/q_n)_{n\geq 0}$ such that the sequence of denominators $(q_n)_{n\geq 0}$ increases without bound and*

$$0 < \left| \alpha - \frac{p_n}{q_n} \right| < \frac{1}{q_n^{1+\epsilon}} \qquad \text{for all} \quad n \in \mathbb{N}. \tag{1.18}$$

Then α is irrational.

First proof By (1.18) the number α is approximable to order $1 + \epsilon$. However, Lemmas 1.32 and 1.33 imply that the best approximation order of a rational number is 1. Therefore α cannot be rational.

Second proof Assume, contrary to what we want to prove, that $\alpha = a/b$ for some $a \in \mathbb{Z}$ and $b \in \mathbb{N}$. In accordance with (1.18) we obtain

$$0 < |bq_n - ap_n| < \frac{b}{q_n^\epsilon} \qquad \text{for all} \quad n \in \mathbb{N}.$$

Since the numbers $bq_n - ap_n$ are *nonzero* integers, we may sharpen the latter inequalities to

$$1 \leq |bq_n - ap_n| < \frac{b}{q_n^\epsilon} \qquad \text{for all} \quad n \in \mathbb{N}. \tag{1.19}$$

Letting $n \to \infty$ in (1.19), we arrive at the inequality $1 \leq 0$. This contradiction shows that $\alpha \in \mathbb{R}$ is irrational. □

The following statement gives a quantitative version of Theorem 1.34, that is, a version which allows one to estimate the irrationality exponent $\mu(\alpha)$ as defined above.

Theorem 1.35 (Quantitative irrationality criterion) *For a given $\alpha \in \mathbb{R}$, assume that there exists an infinite sequence of rational fractions p_n/q_n such that*

$$|q_n\alpha - p_n| = e^{-\psi(n)}, \qquad n = 1, 2, \ldots, \tag{1.20}$$

where the function $\psi : \mathbb{N} \to \mathbb{R}_{>0}$ satisfies the following conditions:

(i) *$\psi(n) \to \infty$ as $n \to \infty$;*

(ii) *$\displaystyle\limsup_{n\to\infty} \frac{\psi(n+1)}{\psi(n)} \leq 1$;*

(iii) *$\displaystyle\rho = \limsup_{n\to\infty} \frac{\log q_n}{\psi(n)} > 0$.*

Then α is irrational, and the estimate $\mu(\alpha) \leq 1 + \rho$ holds.

Proof Let $p \in \mathbb{Z}$ and $q \in \mathbb{N}$ be given with q sufficiently large (its size depends only on the choice of $\epsilon > 0$ below). Define n as smallest positive integer satisfying $2q \leq e^{\psi(n)}$. By using (1.20) we get

$$|(q\alpha - p)q_n| = |q(q_n\alpha - p_n) + (qp_n - pq_n)|$$

$$\geq \begin{cases} 1 - qe^{-\psi(n)} & \text{if } qp_n - pq_n \neq 0; \\ qe^{-\psi(n)} & \text{if } qp_n - pq_n = 0. \end{cases}$$

Hence in both cases we find that

$$\left| \alpha - \frac{p}{q} \right| \geq \frac{1}{q_n e^{\psi(n)}} \geq \frac{1}{e^{(1+\rho+\epsilon)\psi(n)}} \geq \frac{1}{e^{(1+\rho+2\epsilon)\psi(n-1)}},$$

using hypotheses (ii) and (iii). These inequalities yield

$$\left| \alpha - \frac{p}{q} \right| \geq (2q)^{-1-\rho-2\epsilon} > q^{-1-\rho-3\epsilon},$$

and hence $\mu(\alpha) \leq 1 + \rho + 3\epsilon$. But since $\epsilon > 0$ was arbitrary, our claim follows.
□

By Theorem 1.34, condition (i) of Theorem 1.35 is enough to guarantee $\alpha \notin \mathbb{Q}$, whereas (ii) and (iii) are required for the principal quantitative part of the assertion. Condition (ii) means that the function ψ must not grow too quickly – all will be well if everything is polynomial-like.

The approximation orders of different real numbers by rationals are different – and are often not known. Our nearest aim is to show that all irrational numbers from $\mathbb{R}$ are approximable to power order $\nu = 2$; in other words, the irrationality exponent of an irrational number is always at least 2. This implies *a posteriori* that $\rho \geq 1$ in condition (iii) of Theorem 1.35.

DIRICHLET BOX PRINCIPLE   For $m > n$ positive integers, if each of m objects is placed in one of n boxes then there will be at least one box containing more than one object.

In spite of the elementary nature of the *Dirichlet box principle* (also known as the *pigeonhole principle*),[1] found by Peter Gustav Lejeune Dirichlet (1805–1859), it is a useful tool in proving deep results, which are often inaccessible by other methods, not only in number theory but also in other areas in mathematics.

Theorem 1.36 (Dirichlet)   *Let $\alpha \in \mathbb{R}$ and $n \in \mathbb{N}$. Then there exist $p \in \mathbb{Z}$ and $q \in \{1, 2, \ldots, n\}$ such that the following inequality holds:*

$$\left| \alpha - \frac{p}{q} \right| < \frac{1}{qn}. \tag{1.21}$$

---

[1] If there are more letters than letter boxes, someone gets at least two letters.

Proof Consider the $n + 1$ numbers

$$\alpha_k = \{\alpha k\} = \alpha k - \lfloor \alpha k \rfloor, \quad 0 \le \alpha_k < 1, \qquad k = 0, 1, \ldots, n,$$

and subdivide the semi-interval $[0, 1)$ into n equal (nonintersecting) semi-intervals of length $1/n$:

$$[0, 1) = \left[0, \frac{1}{n}\right) \cup \left[\frac{1}{n}, \frac{2}{n}\right) \cup \cdots \cup \left[\frac{n-1}{n}, 1\right).$$

Each of the quantities α_k belongs to exactly one semi-interval. Since the total number of α_k exceeds the number of the semi-intervals by 1, there exists a semi-interval containing at least two distinct quantities, say α_k and α_j, where $0 \le k < j \le n$. Then

$$\frac{1}{n} > |\alpha_j - \alpha_k| = \left|\alpha(j - k) - (\lfloor \alpha j \rfloor - \lfloor \alpha k \rfloor)\right|$$

or, equivalently, in the notation $q = j - k$ and $p = \lfloor \alpha j \rfloor - \lfloor \alpha k \rfloor$,

$$|\alpha q - p| < \frac{1}{n}.$$

It remains to note that $1 \le q \le n$. □

If $\alpha = a/b \in \mathbb{Q}$ with $(a, b) = 1$ then, in view of (1.17), the inequality (1.21) has only the trivial solution $p/q = \alpha$ when $n \ge b$. When $n < b$, Dirichlet's theorem implies that the inequality (1.21) has a solution with denominator $q \le n < b$. Therefore, the denominators of all nontrivial solutions of (1.21) are bounded, and Dirichlet's theorem provides us with certain information on the approximation of rational numbers $\alpha \in \mathbb{Q}$ by rational numbers with smaller denominators.

Lemma 1.37 *If α is irrational then the denominators of the solutions of the inequality* (1.21) *increase with n.*

Proof Assume, contrary to what we want to prove, that there exists an upper bound q_0 for the set of denominators q of the solutions of (1.21). Define the quantities

$$\beta_q = \min_p \left|\alpha - \frac{p}{q}\right|, \quad 1 \le q \le q_0, \qquad \beta = \min_{1 \le q \le q_0} \beta_q;$$

note that $\beta > 0$ by the hypothesis $\alpha \notin \mathbb{Q}$. Then for all $p \in \mathbb{Z}$ and $q \in \mathbb{N}$ we have the inequality

$$\left|\alpha - \frac{p}{q}\right| \ge \beta > 0,$$

which contradicts (1.21) for n sufficiently large. □

The inequality (1.21) and the fact $q \le n$ result in the estimate

$$\left| \alpha - \frac{p}{q} \right| < \frac{1}{q^2}. \tag{1.22}$$

In other words, Lemma 1.37 implies that, in the case of an *irrational* number α, for any $q_0 \in \mathbb{N}$ there exists a solution p/q of (1.22) with $q > q_0$. This demonstrates the following theorem:

Theorem 1.38 (Infinitude of rational approximations)   *For an irrational number α, the inequality (1.22) has infinitely many solutions in $p \in \mathbb{Z}$ and $q \in \mathbb{N}$.*

The theorem implies that all irrational numbers are approximable to power order $\nu = 2$ by rationals.

EXERCISE 1.39 (Chebyshev's theorem [82, Section 8, Theorem 24]; see also Section 7.1 below)   For an irrational number α and a real number β, the inequality

$$|q\alpha + \beta - p| < \frac{3}{q} \tag{1.23}$$

has infinitely many solutions in $p \in \mathbb{Z}$ and $q \in \mathbb{N}$.

EXERCISE 1.40   Show that the number $\sum_{n=0}^{\infty} 2^{-n!}$ is a Liouville number. Show that the set of such numbers is uncountable. (By this method Liouville first exhibited transcendental numbers in 1844, well before Cantor provided cardinality arguments.)

1.5 Farey sequences

Throughout this section, we assume that the denominators of all rational fractions are positive. A fraction a/b is said to be *reducible* if $\gcd(a, b) > 1$; otherwise it is *irreducible*.

The *mediant* of two fractions a/b and $c/d \in \mathbb{Q}$ is the fraction $(a + c)/(b + d)$. In this definition, fractions are not assumed to be irreducible (so that $1/2$ and $2/4$ are here considered as different *fractions*, although they represent the same rational number).

Lemma 1.41   *The mediant of two nonequal fractions coincides with neither of them and always lies between them.*

Proof Without loss of generality we may assume that $a/b < c/d$. Then

$$\frac{a+c}{b+d} - \frac{a}{b} = \frac{bc-ad}{b(b+d)} > 0 \quad \text{and} \quad \frac{a+c}{b+d} - \frac{c}{d} = \frac{ad-bc}{d(b+d)} < 0,$$

since $b > 0$, $d > 0$ and $bc - ad = bd(c/d - a/b) > 0$. $\quad\square$

Two rationals $a/b, c/d \in \mathbb{Q}$ are said to form a *normal pair of fractions* if $bc - ad = 1$.

Lemma 1.42 *If a/b and c/d form a normal pair of fractions then each fraction is irreducible.*

Proof The equality $bc - ad = 1$ yields $\gcd(a,b) = 1$ and $\gcd(c,d) = 1$. $\quad\square$

Lemma 1.43 *If $a/b, c/d$ form a normal pair of fractions then each of the two pairs $a/b, (a+c)/(b+d)$ and $(a+c)/(b+d), c/d$ is a normal pair as well.*

Proof Taking $p = a+c$ and $q = b+d$, direct computation shows that $bp - aq = qc - pd = 1$. $\quad\square$

Lemma 1.44 *Let $a/b, c/d$ be a normal pair of fractions, and let an irreducible fraction p/q satisfy the inequalities*

$$\frac{a}{b} < \frac{p}{q} < \frac{c}{d}.$$

Then $q > b$ and $q > d$.

Proof The hypothesis implies $bc - ad = 1$. Therefore

$$0 < \frac{p}{q} - \frac{a}{b} < \frac{c}{d} - \frac{a}{b} = \frac{1}{bd},$$

and multiplication by bq gives the estimate $0 < bp - aq < q/d$; equivalently, $q/d > bp - aq \geq 1$ or $q > d$. An analogous treatment of the difference $c/d - p/q$ shows that $q > b$. $\quad\square$

Now let $n \in \mathbb{N}$. The *Farey sequence* $\mathcal{F}_n$ *of order* (or *level*) n is the *ordered* (with respect to absolute value) set of the fractions $0/1$, $1/1$ and all positive irreducible regular fractions whose denominators do not exceed n. The first

Farey sequences are as follows:

$$\mathcal{F}_1 : \quad \frac{0}{1}, \frac{1}{1},$$

$$\mathcal{F}_2 : \quad \frac{0}{1}, \frac{1}{2}, \frac{1}{1},$$

$$\mathcal{F}_3 : \quad \frac{0}{1}, \frac{1}{3}, \frac{1}{2}, \frac{2}{3}, \frac{1}{1},$$

$$\mathcal{F}_4 : \quad \frac{0}{1}, \frac{1}{4}, \frac{1}{3}, \frac{1}{2}, \frac{2}{3}, \frac{3}{4}, \frac{1}{1}.$$

If one continues the construction for further values of n using the definition of $\mathcal{F}_n$ then each new step requires more comparisons, so the complexity increases rather quickly.

There are many ways to represent the Farey sequences. One visually attractive way is by drawing *Ford circles* [60], named after Lester Ford. One draws a circle, centred at each rational p/q, with centre $(p/q, 1/(2q^2))$ and radius $1/(2q^2)$. This is illustrated in Figure 1.1. Note that tangent circles are adjacent in the Farey sequence; they are also related to the *Stern–Brocot tree* [68, p. 117].[2]

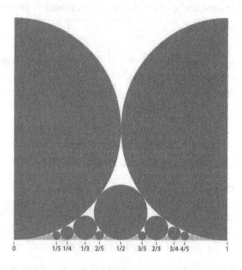

Figure 1.1 Ford circles for F_5: $p/q = 1/2, 1/3, 2/3, 1/4, 3/4, 1/5, 2/5, 3/5, 4/5$.

Let us give a simple and efficient way to construct the Farey sequences. Inductively define an infinite sequence of finite sets $\mathcal{G}_1, \mathcal{G}_2, \ldots, \mathcal{G}_n, \ldots$ of fractions, with the elements in each set ordered by absolute value (as in the Farey

---

[2] See also http://oeis.org/stern_brocot.html.

sequences). The starting set $\mathcal{G}_1$ consists of two fractions: $0/1$ and $1/1$. Further-more, assuming that the set $\mathcal{G}_{n-1}$ is already defined, we construct the set $\mathcal{G}_n$ as follows. In $\mathcal{G}_{n-1}$, run through all adjacent pairs of fractions a/b and c/d with $b + d = n$ and add to $\mathcal{G}_{n-1}$ their mediants. By Lemma 1.41 these mediants are located strictly between the corresponding values of a/b and c/d.

Theorem 1.45 *The (ordered) sets $\mathcal{F}_n$ and $\mathcal{G}_n$ coincide.*

Lemma 1.46 *For each n, any two adjacent fractions in the set $\mathcal{G}_n$ form a normal pair of fractions.*

Proof This is clearly true for $n = 1$. If $n > 1$ then any pair of adjacent fractions in $\mathcal{G}_n$ comes from the set $\mathcal{G}_{n-1}$ in one of the following ways: (1) it is a pair of neighbours already in $\mathcal{G}_{n-1}$, if $b < n$ and $d < n$; (2) a/b is the mediant of neighbours A/B and c/d from $\mathcal{G}_{n-1}$, if $b = n$; or (3) c/d is the mediant of neighbours a/b and C/D from $\mathcal{G}_{n-1}$, if $d = n$. To each of these possibilities we apply the inductive hypothesis and Lemma 1.43. □

Proof of Theorem 1.45 The claim of the theorem holds if $n = 1$. Take $n > 1$. By Lemma 1.46 any two adjacent fractions in $\mathcal{G}_n$ form a normal pair of frac-tions. In particular, by Lemma 1.42, they are irreducible. By the construction, the set $\mathcal{G}_n$ is obtained from $\mathcal{G}_{n-1}$ by adding all mediants of the adjacent frac-tions in $\mathcal{G}_{n-1}$ with denominator sum equal to n. In other words, the denomina-tors of all fractions in $\mathcal{G}_n$ do not exceed n: the set $\mathcal{G}_n$ consists of the fractions $0/1$, $1/1$ and (Lemma 1.41) positive irreducible regular fractions s/t with $t \le n$, that is, $\mathcal{G}_n \subset \mathcal{F}_n$. To prove that $\mathcal{F}_n = \mathcal{G}_n$, we have to show that *all* such fractions s/t with $t \le n$ belong to $\mathcal{G}_n$. Let s/t, where $\gcd(s, t) = 1$ and $0 < s < t \le n$, be any of these fractions and assume, contrary to what we want to prove, that $s/t \notin \mathcal{G}_n$. Take adjacent fractions a/b and c/d in $\mathcal{G}_n$ such that $a/b < s/t < c/d$; note that $b \le n$, $d \le n$ and $b + d \ge n + 1$. By Lemma 1.46, a/b, c/d consti-tute a normal pair of fractions; therefore (by the irreducibility of a mediant) $s/t \ne (a + c)/(b + d)$, since both fractions are irreducible and $t \le n$ while $b + d \ge n + 1$. Thus,

$$\text{either} \quad \frac{a}{b} < \frac{s}{t} < \frac{a+c}{b+d} \quad \text{or} \quad \frac{a+c}{b+d} < \frac{s}{t} < \frac{c}{d}.$$

By Lemma 1.43 each of the pairs of fractions a/b, $(a + c)/(b + d)$ and $(a + c)/(b+d)$, c/d is normal, implying that in each of the above two cases we have $t > b + d \ge n + 1$ by Lemma 1.44, a contradiction (with the hypothesis $t \le n$) completing our proof of the theorem. □

Corollary 1.47 *Any two adjacent fractions in the Farey sequence $\mathcal{F}_n$ of order n form a normal pair of fractions.*

EXERCISE 1.48 How many normal pairs a/b, c/d satisfying $0 < a/b < c/d < 1$ and $\max\{b, d\} \le 10$ exist?

We can now give another proof of Theorem 1.36.

Second proof of Dirichlet's theorem It is sufficient to prove the theorem for $\alpha \in [0, 1)$, since an integer shift of the given α, together with a corresponding shift of the approximation p/q, leads to the same estimate.

Consider the Farey sequence of order n. In $\mathcal{F}_n$ we choose adjacent fractions a/b and c/d such that $a/b \le \alpha < c/d$. Then

$$\text{either} \quad \frac{a}{b} \le \alpha < \frac{a+c}{b+d} \quad \text{or} \quad \frac{a+c}{b+d} \le \alpha < \frac{c}{d}.$$

A property of Farey sequences gives

$$\left| \alpha - \frac{a}{b} \right| < \frac{a+c}{b+d} - \frac{a}{b} = \frac{bc-ad}{b(b+d)} = \frac{1}{b(b+d)}$$

in the first case and

$$\left| \alpha - \frac{c}{d} \right| < \frac{c}{d} - \frac{a+c}{b+d} = \frac{bc-ad}{d(b+d)} = \frac{1}{d(b+d)}$$

in the second. But $b + d \ge n + 1$; hence either

$$\left| \alpha - \frac{a}{b} \right| < \frac{1}{b(n+1)} < \frac{1}{bn}$$

or

$$\left| \alpha - \frac{c}{d} \right| < \frac{1}{d(n+1)} < \frac{1}{dn}.$$

It remains to set $p/q = a/b$ in the first case and $p/q = c/d$ in the second, and the theorem follows. $\square$

REMARK 1.49 We have in fact shown the sharper estimate

$$\left| \alpha - \frac{p}{q} \right| < \frac{1}{q(n+1)}$$

rather than (1.21).

Farey sequences provide an elegant tool for proving diverse deep results in the theory of rational approximations of real numbers. The proofs are based on the fact that one can easily construct a sequence of rationals approaching α sufficiently quickly (that is, what we have just done in the proof above).

Notes

An elementary proof of the prime number theorem was given in the 1940s by Selberg and Erdős; it is, however, technically harder than the original 'advanced' proofs of Hadamard and de la Vallée Poussin. A weaker version of the equivalent form of the prime number theorem given in Theorem 1.20 – namely, the estimate $d_n < 3^n$ – was given by Hanson [71]. This result will be more than sufficient in a further application in Section 9.6.

It follows from Theorem 1.38 that any irrational number α admits a representation of the form

$$\alpha = \frac{p}{q} + \frac{\theta_q}{q^2}, \qquad |\theta_q| < 1, \qquad p \in \mathbb{Z}, \quad q \in \mathbb{N}, \qquad (1.24)$$

where the denominator q is not an arbitrary natural number but can be chosen to be arbitrarily large. The representation in (1.24) possesses some advantages compared with the decimal representation

$$\alpha = a_0 + \frac{a_1}{10} + \frac{a_2}{100} + \cdots + \frac{a_n}{10^n} + \cdots,$$

$$a_0 \in \mathbb{Z}, \quad a_n \in \{0, 1, \ldots, 9\}, \quad n = 1, 2, 3, \ldots.$$

Recall that a 'standard' way to approximate a real number α by rationals is by truncating the decimal record of α,

$$\alpha \approx \frac{A_n}{B_n} = a_0 + \frac{a_1}{10} + \frac{a_2}{100} + \cdots + \frac{a_n}{10^n}, \qquad A_n \in \mathbb{Z}, \quad B_n = 10^n.$$

In this case $\alpha = A_n/B_n + r_n$, where

$$0 < r_n < \frac{9}{10^{n+1}}\left(1 + \frac{1}{10} + \frac{1}{10^2} + \cdots\right) = \frac{1}{10^n} = \frac{1}{B_n},$$

hence

$$0 < \alpha - \frac{A_n}{B_n} < \frac{1}{B_n}.$$

We see that the error in the approximation $\alpha \approx A_n/B_n$ does not exceed $1/B_n$, while the approximation $\alpha \approx p/q$ is of better quality since its inaccuracy is less than $1/q^2$.

Summarising, for any irrational number α there exists a sequence $(p_n/q_n)_{n \geq 1}$ with $q_1 < q_2 < \cdots < q_n < \cdots$ whose terms approximate the number with inaccuracy not exceeding $1/q_n^2$, respectively. Continued fractions, as we shall see in the next chapter, provide us with an efficient way to produce such a sequence of nice rational approximations.

Currently, the best irrationality measure known for π is 7.6063 [145]. For π^2 it is 5.095 412 [176] (see [140] for the previous record), and for $\log 2$ it is 3.574 553 91 [110]. We would be remiss to not mention Roth's theorem that, for any algebraic number α,

$$\left| \alpha - \frac{p}{q} \right| < \frac{1}{q^{2+\varepsilon}} \tag{1.25}$$

with $\varepsilon > 0$ has only finitely many solutions. Thus, if there are infinitely many solutions to (1.25) then the number is transcendental, while usually having a finite irrationality measure unlike the case of Liouville numbers.

Farey sequences give rise to some elementary formulations of the Riemann hypothesis. These are based on measuring how uniformly the Farey fractions are distributed. To make this more precise, consider the Farey sequence of order n,

$$\mathcal{F}_n = \left\{ \frac{0}{1} = \frac{a_0}{b_0} < \frac{a_1}{b_1} < \cdots < \frac{a_m}{b_m} = \frac{1}{1} \right\},$$

where of course $m = m_n$ is an increasing function in n; form the deviations

$$d_i = d_{i,n} = \left| \frac{a_i}{b_i} - \frac{i}{m} \right|, \quad i = 0, 1, \ldots, m,$$

from the uniform distribution on the unit segment $[0, 1]$. Franel [64] and Landau [93] proved that the two statements

$$\sum_{i=1}^{m_n} d_{i,n} = O(n^\delta) \quad \text{for any } \delta > 1/2$$

and

$$\sum_{i=1}^{m_n} d_{i,n}^2 = O(n^\delta) \quad \text{for any } \delta > -1$$

are each equivalent to the Riemann hypothesis. These and many other equivalences can be followed up in [32].

2

Continued fractions

We are now ready to begin our study of continued fractions.

2.1 A π-overview of the theory

In the years BC – before calculators – π was $22/7$ and in the years AD – after decimals – π became $3.141\,592\,65\ldots$ Clearly, π is reasonably well approximated by the vulgar fraction $22/7$, and some of us know that $355/113$ does a yet better job since it yields as many as seven correct decimal digits.

The 'why this is so' of the matter is as follows. It happens that

$$\pi = 3 + \cfrac{1}{7 + \cfrac{1}{15 + \cfrac{1}{1 + \cfrac{1}{292 + \cfrac{1}{1 + \ddots}}}}} \qquad (2.1)$$

and, in particular, that

$$22/7 = 3 + \frac{1}{7} \qquad \text{while} \qquad 355/113 = 3 + \cfrac{1}{7 + \cfrac{1}{15 + \cfrac{1}{1}}}.$$

Obviously, the notation takes too much space. We also note that *truncations of the neverending fraction* (2.1) *seem to provide very good rational approximations.* (We have coined the term *neverending fraction* as a synonym for an infinite continued fraction, which is what equations like (2.1) represent. We will of course need to consider convergence issues.)

In example (2.1) it is only the *partial quotients* $3, 7, 15, \ldots$ that matter, so we may conveniently write (2.1) as

$$\pi = [3; 7, 15, 1, 292, 1, \ldots].$$

In general, given an irrational number α, define its sequence $\alpha_0, \alpha_1, \alpha_2, \ldots$ of *complete quotients* by setting $\alpha_0 = \alpha$ and $\alpha_{n+1} = 1/(\alpha_n - a_n)$. Here, the sequence $a_0, a_1, a_2, \ldots$ of *partial quotients* of α is given by $a_n = \lfloor \alpha_n \rfloor$, where $\lfloor \cdot \rfloor$ is the greatest integer function, introduced in Section 1.1. The truncations $[a_0; a_1, \ldots, a_n]$ are plainly rational numbers p_n/q_n. Indeed, the *continuants* p_n and q_n are defined by the matrix identities

$$\begin{pmatrix} a_0 & 1 \\ 1 & 0 \end{pmatrix} \begin{pmatrix} a_1 & 1 \\ 1 & 0 \end{pmatrix} \cdots \begin{pmatrix} a_n & 1 \\ 1 & 0 \end{pmatrix} = \begin{pmatrix} p_n & p_{n-1} \\ q_n & q_{n-1} \end{pmatrix}, \qquad n = 1, 2, \ldots$$

This follows readily by induction on n and the definition

$$[a_0; a_1, \ldots, a_n] = a_0 + \frac{1}{[a_1; \ldots, a_n]}, \qquad [a_0] = a_0.$$

Taking determinants in the matrix correspondence above immediately implies that the *convergents* p_n/q_n satisfy

$$\frac{p_n}{q_n} - \frac{p_{n-1}}{q_{n-1}} = \frac{(-1)^{n-1}}{q_{n-1} q_n}, \qquad (2.2)$$

and so

$$\frac{p_n}{q_n} = a_0 + \sum_{j=1}^{n-1} \frac{(-1)^{j-1}}{q_{j-1} q_j},$$

showing that the convergents do converge to a limit, namely

$$\alpha = a_0 + \sum_{n=1}^{\infty} \frac{(-1)^{n-1}}{q_{n-1} q_n}, \qquad (2.3)$$

and also that

$$0 < (-1)^{n-1} \left(\alpha - \frac{p_n}{q_n} \right) < \frac{1}{q_n q_{n+1}} < \frac{1}{a_{n+1} q_n^2}.$$

REMARK 2.1   The use of matrix notation often dramatically simplifies proofs, as is the case with (2.2).

Thus, in particular

$$0 < \frac{22}{7} - \pi < \frac{1}{15 \times 7^2} \qquad \text{and} \qquad 0 < \frac{355}{113} - \pi < \frac{1}{292 \times 113^2}. \qquad (2.4)$$

Conversely, suppose that $q_{n-1} < q < q_n$. Because $\gcd(q_{n-1}, q_n) = 1$ there are integers a and b, with $ab < 0$, such that $q = a q_{n-1} + b q_n$ (see Exercise 1.12).

Set $p = ap_{n-1} + bp_n$. Then $q\alpha - p$ is $a(q_{n-1}\alpha - p_{n-1}) + b(q_n\alpha - p_n)$ and, *since the two terms have the same sign*, each must be smaller than $|q\alpha - p|$ in absolute value. Thus convergents yield *locally best approximations* and it follows that certainly $q\alpha - p > 1/(2q)$.

EXERCISE 2.2 (See [51], [22, Section 1.1] and [105]) Verify that

$$\int_0^1 \frac{x^4(1-x)^4}{1+x^2}\, dx = \frac{22}{7} - \pi,$$

$$\int_0^1 \frac{x^8(1-x)^8(25 + 816x^2)}{3164(1+x^2)}\, dx = \frac{355}{113} - \pi,$$

and use the integrals to give an alternative proof of the estimates (2.4). Note that the integrands are strictly positive on the open interval and hence the integral is strictly positive.

It appears to be largely coincidental that this works for the two early convergents of π. That is, we know of no methodical way to undertake such a process of representing continued fraction errors by integrals of positive rational functions.

2.2 Finite continued fractions and the Euclidean algorithm

A *finite continued fraction* is the expression

$$[a_0; a_1, a_2, \ldots, a_n] = a_0 + \cfrac{1}{a_1 + \cfrac{1}{a_2 + \cfrac{1}{\ddots + \cfrac{1}{a_n}}}}, \tag{2.5}$$

where $a_0 \in \mathbb{Z}$ and $a_1, \ldots, a_n \in \mathbb{N}$ are the *partial quotients* of the continued fraction.

As a result of the finite number of arithmetic operations involved in its definition, a finite continued fraction is clearly a rational number. This number is called the *value* of the continued fraction.

EXERCISE 2.3 Find the value of the finite continued fraction $[1; 2, 3, 4, 5]$.

Theorem 2.4 *Any rational number admits a representation in the form of continued fraction* (2.5) *with last term $a_n > 1$ if $n \geq 1$.*

We shall discuss the uniqueness of such representations later.

Proof Let a/b be a rational number with $a \in \mathbb{Z}$, $b \in \mathbb{N}$, and $(a, b) = 1$. Setting $r_{-1} = a$ and $r_0 = b$, write the Euclidean algorithm (successive application of division with remainder) as

$$r_{-1} = r_0 a_0 + r_1, \quad r_0 = r_1 a_1 + r_2, \quad r_1 = r_2 a_2 + r_3, \quad \ldots,$$
$$r_{n-2} = r_{n-1} a_{n-1} + r_n, \quad r_{n-1} = r_n a_n, \tag{2.6}$$

where we have

$$0 < r_n < r_{n-1} < \cdots < r_0 = b. \tag{2.7}$$

If $b = 1$ then the number in question, a/b, is an integer, and hence the sequence (2.6) consists of the single equality $a = r_{-1} = 1 \times a_0 + 0$ only, so that the continued fraction expansion of a/b takes the form $a = [a_0] = [a]$. Otherwise, if $b > 1$ then a is not divisible by b, since $(a, b) = 1$. Therefore, $r_1 > 0$ and the sequence (2.6) consists of more than one iteration.

It follows from the first equality that $a < 0$ implies $a_0 < 0$, while if $a > 0$ and the fraction a/b is regular (that is, $a < b$) then $a_0 = 0$. As for the other equalities in (2.6), by the inequalities (2.7) we deduce that $a_1, \ldots, a_n \in \mathbb{N}$ and $a_n > 1$ (the latter follows from the equality $r_{n-1} = r_n a_n$ and the inequality $r_n < r_{n-1}$). Dividing both sides of each equality in (2.6) by $r_0, r_1, \ldots, r_{n-1}, r_n$, respectively, we get

$$\frac{a}{b} = \frac{r_{-1}}{r_0} = a_0 + \frac{r_1}{r_0} = a_0 + \frac{1}{r_0/r_1},$$
$$\frac{r_0}{r_1} = a_1 + \frac{r_2}{r_1} = a_1 + \frac{1}{r_1/r_2},$$
$$\frac{r_1}{r_2} = a_2 + \frac{r_3}{r_2} = a_2 + \frac{1}{r_2/r_3},$$
$$\vdots$$
$$\frac{r_{n-2}}{r_{n-1}} = a_{n-1} + \frac{r_n}{r_{n-1}} = a_{n-1} + \frac{1}{r_{n-1}/r_n},$$
$$\frac{r_{n-1}}{r_n} = a_n.$$

Starting from the last equality and substituting each equality into the previous one, we finally arrive at the desired expansion (2.5). □

We now see that the terms $a_0, a_1, \ldots, a_n$ are called *partial quotients* because the successive divisions in the Euclidean algorithm involve these numbers as partial quotients of the corresponding quantities.

EXERCISE 2.5 (The Euclid game [39]) The game *Euclid* is played with a pair of nonnegative integers. Two players move alternately, each subtracting a positive integer multiple of one integer from the other integer without making the result negative. The player who reduces one of the integers to zero wins.

In a subtle variation, the game stops when the two numbers are equal. When does a winning strategy exist for the first player?

2.3 Algebraic theory of continued fractions

Let us now take a slightly different approach. Viewing $a_0, a_1, \ldots, a_n, \ldots$ as independent variables, define consecutively pairs of polynomials in these variables

$$p_n = p_n(a_0; a_1, \ldots, a_n), \qquad q_n = q_n(a_0; a_1, \ldots, a_n)$$

by setting $p_0 = a_0$, $q_0 = 1$ and

$$
\begin{aligned}
p_n &= a_0 p_{n-1}(a_1; a_2, \ldots, a_n) + q_{n-1}(a_1; a_2, \ldots, a_n), \\
q_n &= p_{n-1}(a_1; a_2, \ldots, a_n).
\end{aligned}
\tag{2.8}
$$

We now introduce a convenient notation for the quotient of the polynomials:

$$\frac{p_n}{q_n} = [a_0; a_1, \ldots, a_n].$$

Then

$$[a_0; a_1, \ldots, a_n] = a_0 + \frac{1}{[a_1; a_2, \ldots, a_n]}.
\tag{2.9}$$

Developing this equality further we obtain

$$
\begin{aligned}
[a_0; a_1, \ldots, a_n] &= a_0 + \cfrac{1}{a_1 + \cfrac{1}{[a_2; a_3, \ldots, a_n]}} \\
&= a_0 + \cfrac{1}{a_1 + \cfrac{1}{a_2 + \cfrac{1}{a_3 + \cfrac{}{\ddots + \cfrac{1}{a_n}}}}}.
\end{aligned}
\tag{2.10}
$$

This is the expression we have agreed to call a finite continued fraction.

Example 2.6 Using the above definition we find that

$$p_1 = a_0 a_1 + 1 \quad \text{and} \quad q_1 = a_1.$$

REMARK 2.7   In fact, it is often helpful to define the polynomials p_n and q_n in accordance with (2.8) but starting from the initial data $p_{-1} = 1$ and $q_{-1} = 0$ for $n = -1$ (when the set of variables of the corresponding polynomials is empty). Then $p_0 = a_0$ and $q_0 = 1$ by (2.8), and this agrees with the initial data above.

Summarising the construction above, we associate the sequence of continued fractions (2.10) with a set of independent variables $a_0, a_1, \ldots, a_n, \ldots$. If we replace the variables $a_0, a_1, \ldots, a_n, \ldots$ with (not necessarily integer) real numbers such that the values of q_n do not vanish then we get a sequence of numbers written in the form of continued fractions.

A simpler (and, in a certain sense, more natural) way of writing the paired equations (2.8) is the matrix form

$$\begin{pmatrix} p_n \\ q_n \end{pmatrix} = \begin{pmatrix} a_0 & 1 \\ 1 & 0 \end{pmatrix} \begin{pmatrix} p'_{n-1} \\ q'_{n-1} \end{pmatrix}, \tag{2.11}$$

where the prime means that the set of variables is shifted by one:

$$p'_{n-1} = p_{n-1}(a_1; a_2, \ldots, a_n) \quad \text{and} \quad q'_{n-1} = q_{n-1}(a_1; a_2, \ldots, a_n).$$

Iterating the matrix relation (2.11) we obtain

$$\begin{pmatrix} p_n \\ q_n \end{pmatrix} = \begin{pmatrix} a_0 & 1 \\ 1 & 0 \end{pmatrix} \begin{pmatrix} a_1 & 1 \\ 1 & 0 \end{pmatrix} \cdots \begin{pmatrix} a_{n-1} & 1 \\ 1 & 0 \end{pmatrix} \begin{pmatrix} a_n \\ 1 \end{pmatrix} \tag{2.12}$$

$$= \begin{pmatrix} a_0 & 1 \\ 1 & 0 \end{pmatrix} \begin{pmatrix} a_1 & 1 \\ 1 & 0 \end{pmatrix} \cdots \begin{pmatrix} a_{n-1} & 1 \\ 1 & 0 \end{pmatrix} \begin{pmatrix} a_n & 1 \\ 1 & 0 \end{pmatrix} \begin{pmatrix} 1 \\ 0 \end{pmatrix}. \tag{2.13}$$

Combining the representations (2.12) for n and (2.13) for $n - 1$, we arrive at

Lemma 2.8 (Key lemma)   *For each $n \geq 1$, we have the relations*

$$\begin{pmatrix} p_n & p_{n-1} \\ q_n & q_{n-1} \end{pmatrix} = \begin{pmatrix} a_0 & 1 \\ 1 & 0 \end{pmatrix} \begin{pmatrix} a_1 & 1 \\ 1 & 0 \end{pmatrix} \cdots \begin{pmatrix} a_{n-1} & 1 \\ 1 & 0 \end{pmatrix} \begin{pmatrix} a_n & 1 \\ 1 & 0 \end{pmatrix} = \prod_{j=0}^{n} \begin{pmatrix} a_j & 1 \\ 1 & 0 \end{pmatrix}. \tag{2.14}$$

The matrix form (2.14) is a very useful basis for proving practically everything about the formal continued fraction (2.9).

Lemma 2.9   *For each $n \geq 1$, we have the relations*

$$p_n = a_n p_{n-1} + p_{n-2}, \qquad q_n = a_n q_{n-1} + q_{n-2}. \tag{2.15}$$

Proof   This follows immediately from

$$\begin{pmatrix} p_n & p_{n-1} \\ q_n & q_{n-1} \end{pmatrix} = \begin{pmatrix} p_{n-1} & p_{n-2} \\ q_{n-1} & q_{n-2} \end{pmatrix} \begin{pmatrix} a_n & 1 \\ 1 & 0 \end{pmatrix}. \qquad \square$$

EXERCISE 2.10 (Repeated 1s)   Find the value of the finite continued fraction

$$[\{1\}^n] = \underbrace{[1; 1, 1, \ldots, 1]}_{n \text{ times}}$$

and compute the limit $[\{1\}^\infty] = \lim_{n \to \infty} [\{1\}^n]$.

At present we are interested in the case when $a_1, a_2, \ldots$ are positive real numbers; their positivity guarantees that $q_n > 0$ for $n \geq 0$. Then the fraction p_n/q_n is a certain real number, and the definition (2.10) of a continued fraction together with Lemma 2.9 imply the following statements.

Lemma 2.11   *Let a_i, $i = 1, \ldots, n$, be positive reals. Define the complete quotients (the 'tails')*

$$r_k = [a_k; a_{k+1}, \ldots, a_n], \qquad k = 1, \ldots, n.$$

Then

$$[a_0; a_1, \ldots, a_n] = [a_0; a_1, \ldots, a_{k-1}, r_k] = \frac{p_{k-1} r_k + p_{k-2}}{q_{k-1} r_k + q_{k-2}}.$$

Proof   This statement is nothing other than the matrix identity

$$\prod_{j=0}^{n} \begin{pmatrix} a_j & 1 \\ 1 & 0 \end{pmatrix} = \prod_{j=0}^{k-1} \begin{pmatrix} a_j & 1 \\ 1 & 0 \end{pmatrix} \times \prod_{j=k}^{n} \begin{pmatrix} a_j & 1 \\ 1 & 0 \end{pmatrix}.$$

Namely, we have $r_k = p'_k / q'_k$, where

$$\begin{pmatrix} p'_k \\ q'_k \end{pmatrix} = \begin{pmatrix} a_k & 1 \\ 1 & 0 \end{pmatrix} \begin{pmatrix} a_{k+1} & 1 \\ 1 & 0 \end{pmatrix} \cdots \begin{pmatrix} a_n & 1 \\ 1 & 0 \end{pmatrix} \begin{pmatrix} 1 \\ 0 \end{pmatrix};$$

then

$$\begin{pmatrix} p_n \\ q_n \end{pmatrix} = \begin{pmatrix} a_n & 1 \\ 1 & 0 \end{pmatrix} \cdots \begin{pmatrix} a_{k-1} & 1 \\ 1 & 0 \end{pmatrix} \begin{pmatrix} p'_k \\ q'_k \end{pmatrix} = \begin{pmatrix} p_{k-1} & p_{k-2} \\ q_{k-1} & q_{k-2} \end{pmatrix} \begin{pmatrix} p'_k \\ q'_k \end{pmatrix}$$

$$= \begin{pmatrix} p_{k-1} p'_k + p_{k-2} q'_k \\ q_{k-1} p'_k + q_{k-2} q'_k \end{pmatrix};$$

hence

$$[a_0; a_1, \ldots, a_n] = \frac{p_n}{q_n} = \frac{p_{k-1} p'_k + p_{k-2} q'_k}{q_{k-1} p'_k + q_{k-2} q'_k} = \frac{p_{k-1} r_k + p_{k-2}}{q_{k-1} r_k + q_{k-2}}. \qquad \square$$

Lemma 2.12   *Let $a_0, a_1, \ldots, a_n$ and $b_0, b_1, \ldots, b_n$ be positive real numbers satisfying $a_i \geq 1$ and $b_i \geq 1$ for $i = 1, \ldots, n$. Assume that the numbers a_i, b_i are integers for $i = 0, \ldots, n - 1$. Then the equality*

$$[a_0; a_1, \ldots, a_n] = [b_0; b_1, \ldots, b_n]$$

implies that $a_i = b_i$ for all $i = 0, 1, \ldots, n$.

Proof For

$$r_1 = [a_1; a_2, \ldots, a_n] = a_1 + \cfrac{1}{[a_2; a_3, \ldots, a_n]}$$

we have $r_1 \geq 1$, and the same reasoning implies that $s_1 = [b_1; b_2, \ldots, b_n] \geq 1$. By the hypothesis,

$$a_0 + \frac{1}{r_1} = b_0 + \frac{1}{s_1}. \tag{2.16}$$

If $r_1 = 1$ then the left-hand side of (2.16) is an integer. Hence $1/s_1$ is an integer as well, implying $s_1 = 1$ and $b_0 = a_0$. Otherwise, if $r_1 > 1$ then the left-hand side of (2.16) is not in $\mathbb{Z}$ and we deduce that $s_1 > 1$ and, consequently, $a_0 = b_0$, since both a_0 and b_0 are the largest integers which do not exceed $[a_0; a_1, \ldots, a_n] = [b_0; b_1, \ldots, b_n]$. In all cases we get $a_0 = b_0$ hence $r_1 = s_1$. Induction on n completes the proof of the statement.   □

Lemma 2.12 actually shows the uniqueness of the continued fraction representation (2.10) (with $a_n > 1$ if $n \geq 1$) of a *rational* number; the existence was proved in Theorem 2.4.

EXERCISE 2.13 Deduce the uniqueness of the representation from Theorem 2.4.

2.4 Relations for continuants of a continued fraction

In the statements in this section we consider $a_0, a_1, \ldots, a_n, \ldots$ as indeterminates, which may even lie in function fields.

Theorem 2.14 *For $n \geq 0$, the relation*

$$q_n p_{n-1} - p_n q_{n-1} = (-1)^n$$

holds.

Proof This is the determinant evaluation of (2.14) after noting that

$$\det \begin{pmatrix} a_j & 1 \\ 1 & 0 \end{pmatrix} = -1.$$ □

Corollary 2.15 *For $n \geq 1$, the following relation holds:*

$$\frac{p_{n-1}}{q_{n-1}} - \frac{p_n}{q_n} = \frac{(-1)^n}{q_{n-1} q_n}.$$

Corollary 2.16 *If $a_0, a_1, a_2, \ldots$ are positive integers then p_n and q_n are coprime and $0 < q_1 < q_2 < \cdots < q_n < \cdots$, that is, the denominators q_n form a strictly increasing sequence of integers.*

Corollary 2.17 *Consider* $\alpha = [a_0; a_1, \ldots, a_{n+2}]$ *as a rational function of the variables* $a_0, a_1, \ldots, a_{n+2}$. *Then*

$$q_{n+1}\alpha - p_{n+1} = \frac{(-1)^{n+1}}{a_{n+2}q_{n+1} + q_n}.$$

Proof Replace n in Theorem 2.14 with $n + 2$ and divide both sides of the corresponding relation by q_{n+2}. Noting that $p_{n+2}/q_{n+2} = \alpha$ by definition and that $q_{n+2} = a_{n+2}q_{n+1} + q_n$ by Lemma 2.9, we deduce the desired relation. $\square$

Theorem 2.18 *For* $n \geq 1$, *the relation*

$$q_n p_{n-2} - p_n q_{n-2} = (-1)^{n-1} a_n$$

holds.

Proof To prove this relation, we can obtain from (2.13) the matrix identity

$$\begin{pmatrix} p_n & p_{n-2} \\ q_n & q_{n-2} \end{pmatrix} = \begin{pmatrix} a_0 & 1 \\ 1 & 0 \end{pmatrix}\begin{pmatrix} a_1 & 1 \\ 1 & 0 \end{pmatrix} \cdots \begin{pmatrix} a_{n-2} & 1 \\ 1 & 0 \end{pmatrix}\begin{pmatrix} a_{n-1}a_n + 1 & 1 \\ a_n & 0 \end{pmatrix}$$

and then compute the determinants of both sides.

Alternatively, multiply the first expression in Lemma 2.9 by q_{n-2} and the second by p_{n-2} and subtract the first equality from the second. In accordance with Theorem 2.14 we get

$$q_n p_{n-2} - p_n q_{n-2} = a_n(q_{n-1}p_{n-2} - p_{n-1}q_{n-2}) = (-1)^{n-1} a_n,$$

finishing the proof. $\square$

Corollary 2.19 *For* $n \geq 2$ *we have*

$$\frac{p_{n-2}}{q_{n-2}} - \frac{p_n}{q_n} = \frac{(-1)^{n-1} a_n}{q_{n-2}q_n}.$$

Corollary 2.20 *If* $a_1, a_2, \ldots$ *are positive (not necessarily integer) numbers then the sequence* p_n/q_n *restricted to even* n *is strictly increasing, while restricted to odd* n *it is strictly decreasing.*

Corollary 2.21 *Consider* $\alpha = [a_0; a_1, \ldots, a_{n+2}]$ *as a rational function of the variables* $a_0, a_1, \ldots, a_{n+2}$. *Then*

$$q_n\alpha - p_n = \frac{(-1)^n a_{n+2}}{a_{n+2}q_{n+1} + q_n}.$$

Proof Replace n in Theorem 2.18 with $n + 2$ and divide both sides of the formula by q_{n+2}. Note that $p_{n+2}/q_{n+2} = \alpha$ by definition and $q_{n+2} = a_{n+2}q_{n+1} + q_n$ by Lemma 2.9. This implies the desired statement. $\square$

EXERCISE 2.22   Prove the following identities for $n \geq 1$:

(a) $q_n/q_{n-1} = [a_n; a_{n-1}, \ldots, a_1]$;
(b) $p_n/p_{n-1} = [a_n; a_{n-1}, \ldots, a_1, a_0]$.

Solution   Transpose the matrix identity (2.14).   □

EXERCISE 2.23 (H. J. S. Smith; see [46])   For continuants $p_n = p_n(a_0; a_1, \ldots, a_n)$ and $q_n = q_n(a_0; a_1, \ldots, a_n)$ show the following determinant expressions:

$$
p_n = \det \begin{pmatrix}
a_0 & 1 & 0 & \cdots & 0 & 0 \\
-1 & a_1 & 1 & \cdots & 0 & 0 \\
0 & -1 & a_2 & \cdots & 0 & 0 \\
\vdots & \vdots & \vdots & \ddots & \vdots & \vdots \\
0 & 0 & \cdots & \cdots & a_{n-1} & 1 \\
0 & 0 & \cdots & \cdots & -1 & a_n
\end{pmatrix},
$$

$$
q_n = \det \begin{pmatrix}
a_1 & 1 & 0 & \cdots & 0 & 0 \\
-1 & a_2 & 1 & \cdots & 0 & 0 \\
0 & -1 & a_3 & \cdots & 0 & 0 \\
\vdots & \vdots & \vdots & \ddots & \vdots & \vdots \\
0 & 0 & \cdots & \cdots & a_{n-1} & 1 \\
0 & 0 & \cdots & \cdots & -1 & a_n
\end{pmatrix}.
$$

2.5 Continued fraction of a real number

We have already seen, in Section 2.2, that the (finite) continued fraction of a rational number α can be found using the Euclidean algorithm (Theorem 2.4): take $a_0 = \lfloor \alpha \rfloor$ and, if α is not an integer, then write it in the form $\alpha = a_0 + 1/\alpha_1$, where $\alpha_1 > 1$ is again a rational number. Inductively, we choose $a_n = \lfloor \alpha_n \rfloor$ and $\alpha_n = a_n + 1/\alpha_{n+1}$ with $\alpha_{n+1} > 1$ if α_n is not an integer. The procedure terminates at some step (that is, eventually we get an integer $\alpha_n = a_n > 1$), so that $\alpha = [a_0; a_1, \ldots, a_n]$. If we discard the condition $a_n > 1$ for $n \geq 1$ then the number α can be also represented as $\alpha = [a_0; a_1, \ldots, a_n - 1, 1]$. This fact is sometimes useful for manipulating the parity of a particular length of a finite continued fraction and for other similar reasons.

The recursive algorithm above extends to the case of an *irrational* number α with no trouble; however, at each step we obtain irrational $\alpha_n > 1$, so that the continued fraction cannot be finite. We will use the notation

$$\alpha = [a_0; a_1, \ldots, a_n, \ldots]$$

for this infinite case.

From Corollary 2.16 we obtain a sequence of coprime integers p_n and q_n, defined as the numerator and denominator of each finite continued fraction $[a_0; a_1, \ldots, a_n]$; in addition, $0 < q_1 < q_2 < \cdots < q_n < \cdots$. In this case, the equality

$$\frac{p_n}{q_n} = [a_0; a_1, \ldots, a_n]$$

becomes a relation between real numbers rather than between functions involving indeterminates $a_0, a_1, \ldots$ Moreover p_n/q_n is an irreducible fraction, which is called the *n*th (*principal*) *convergent* of α. The number a_n is called the *n*th *partial quotient* of the number α. In this notation, the statements of Corollaries 2.17 and 2.21 (after replacement of a_{n+2} with α_{n+2}) take the following form.

Lemma 2.24 *For $n \geq 0$, we have the equalities*

$$q_{n+1}\alpha - p_{n+1} = \frac{(-1)^{n+1}}{\alpha_{n+2}q_{n+1} + q_n}, \qquad q_n\alpha - p_n = \frac{(-1)^n \alpha_{n+2}}{\alpha_{n+2}q_{n+1} + q_n}.$$

We stress that $1 \leq a_n < \alpha_n < a_n + 1$ for all $n \geq 1$.

Theorem 2.25 (Monotonicity and estimation of convergents) *For even n, the nth convergents of α form a strictly increasing sequence converging to α; for odd n, the nth convergents of α form a strictly decreasing sequence converging to α. Furthermore,*

$$\frac{1}{2q_nq_{n+1}} < \frac{1}{q_n(q_n + q_{n+1})} < \left| \alpha - \frac{p_n}{q_n} \right| < \frac{1}{q_nq_{n+1}} < \frac{1}{a_{n+1}q_n^2}.$$

Proof The first part of the claim follows from Corollaries 2.20 and 2.15; Corollary 2.15 also justifies the estimate from above for $|\alpha - p_n/q_n|$. To show the validity of the estimate from below we use Corollary 2.19:

$$\left| \alpha - \frac{p_n}{q_n} \right| > \left| \frac{p_{n+2}}{q_{n+2}} - \frac{p_n}{q_n} \right| = \frac{a_{n+2}}{q_nq_{n+2}} = \frac{a_{n+2}}{q_n(a_{n+2}q_{n+1} + q_n)}.$$

On dividing the numerator and denominator of the last fraction by $a_{n+2} \geq 1$, the theorem follows. □

Since $q_{n+1} > q_n$, we conclude that the principal convergents p_n/q_n satisfy the inequality

$$\left| \alpha - \frac{p}{q} \right| < \frac{1}{q^2}.$$

EXERCISE 2.26 (Dirichlet's theorem) Using continued fractions, give a third proof of Dirichlet's theorem (Theorem 1.36) for an irrational real number α.

Hint For a given integer $n > 0$, take the index j such that $q_j < n \le q_{j+1}$ and show that the convergent $p/q = p_j/q_j$ satisfies

$$\left| \alpha - \frac{p}{q} \right| < \frac{1}{qn}. \qquad \qquad \square$$

EXERCISE 2.27 (Sandwiching principle) Given a positive real number α, show that if a/b and c/d are positive rational numbers such that

$$\frac{a}{b} < \alpha < \frac{c}{d}$$

then, as long as the partial quotients of a/b and c/d coincide, those partial quotients are the partial quotients of α. The first time the partial quotients do not coincide, they provide upper and lower bounds for the correct value.

There is in fact no reason to restrict ourselves to the *integer* specialisation of partial quotients $a_0, a_1, \ldots$

EXERCISE 2.28 (Continued fractions over fields) For a field K ($\mathbb{Q}$, $\mathbb{R}$, $\mathbb{C}$ or a finite field), show that a continued fraction with partial quotients in $K[x]$ converges to a Laurent series in $K((1/x))$.

EXERCISE 2.29 Define $S_0 = 2$ and $S_{n+1} = S_n^2 - S_n + 1$ for $n \ge 0$; this is Sylvester's sequence.

(a) Using (2.3) show that the partial quotients in the continued fraction expansion of

$$C = \sum_{i=0}^{\infty} \frac{(-1)^i}{S_i - 1}$$

are all squares.

(b) Prove that

$$C' = \sum_{i=0}^{\infty} \frac{(-1)^i}{S_i}$$

has a continued fraction $[a_0; a_1, \ldots, a_k, \ldots]$ where for $k \ge 2$ each $a_k/2$ is a square.

(c) Finally, show that $2C = C' + 1$.

The number $C = 0.643\,410\,546\,28\ldots$ is sometimes called *Cahen's constant*; see [56] and [58, Section 6.7].

2.6 A taste of Diophantine approximation

Lemma 2.30 *Let $\alpha = [a_0, a_1, a_2, \dots]$ be a continued fraction and p_n/q_n, $n \geq 0$, the sequence of its convergents. Then the inequalities*

$$\frac{1}{(a_{n+1} + 2)q_n^2} < \left| \alpha - \frac{p_n}{q_n} \right| < \frac{1}{a_{n+1}q_n^2}, \qquad n \geq 1, \tag{2.17}$$

$$a_1 \cdots a_n \leq q_n \leq F_n a_1 \cdots a_n, \qquad n \geq 1, \tag{2.18}$$

hold, where F_n denotes the Fibonacci sequence (defined in Section 1.3).

Proof By Lemma 2.24 we have the equality

$$\left| \alpha - \frac{p_n}{q_n} \right| = \frac{1}{(\alpha_{n+1}q_n + q_{n-1})q_n},$$

where $\alpha_{n+1} = [a_{n+1}, a_{n+2}, \dots]$ is the complete quotient of the continued fraction. Therefore, the inequalities $a_{n+1} \leq \alpha_{n+1} < a_{n+1} + 1$ and $0 \leq q_{n-1} \leq q_n$ imply the estimates (2.17).

To prove (2.18), we proceed by induction. The inequalities (2.18) are clearly true for $n = 1$ and $n = 2$, while for $n \geq 2$ the equality $q_{n+1} = a_{n+1}q_n + q_{n-1}$ implies $a_{n+1}q_n \leq q_{n+1} \leq a_{n+1}(q_n + q_{n-1})$. The latter inequalities together with the inductive hypothesis and the recurrence relation $F_{n+1} = F_n + F_{n-1}$ for the Fibonacci numbers result in (2.18). $\square$

EXERCISE 2.31 (Repeated 1s; see Exercise 2.10)

(a) Show that $[\{1\}^\infty] = (1 + \sqrt{5})/2$. Thus the *golden mean* (or *ratio*), φ, yields the most basic periodic continued fraction.

(b) Evaluate $[\{k\}^\infty]$ for $2 \leq k \leq 10$.

Lemma 2.32 *Let p_{n-1}/q_{n-1} and p_n/q_n be two successive convergents of an irrational number $\alpha = [a_0; a_1, a_2, \dots]$. Then at least one of these fractions satisfies the inequality*

$$\left| \alpha - \frac{p}{q} \right| < \frac{1}{2q^2}.$$

Proof Assume, to get a contradiction, that

$$\left| \alpha - \frac{p_{n-1}}{q_{n-1}} \right| \geq \frac{1}{2q_{n-1}^2} \qquad \text{and} \qquad \left| \alpha - \frac{p_n}{q_n} \right| \geq \frac{1}{2q_n^2}.$$

Using the fact that α lies between p_{n-1}/q_{n-1} and p_n/q_n (Theorem 2.25) we obtain

$$\frac{1}{q_{n-1}q_n} = \left| \frac{p_{n-1}}{q_{n-1}} - \frac{p_n}{q_n} \right| = \left| \alpha - \frac{p_{n-1}}{q_{n-1}} \right| + \left| \alpha - \frac{p_n}{q_n} \right| \geq \frac{1}{2q_{n-1}^2} + \frac{1}{2q_n^2}.$$

This contradicts the inequality $xy < (x^2 + y^2)/2$ for $x > y > 0$, applied with $x = 1/q_{n-1}$ and $y = 1/q_n$. □

Our next statement shows that, in a certain sense, the converse of Lemma 2.32 holds as well.

Lemma 2.33 (Legendre's theorem) *Let p and q be coprime integers, $q > 0$, and let*

$$\left| \alpha - \frac{p}{q} \right| < \frac{1}{2q^2}.$$

Then p/q is a convergent of α.

Proof Write the continued fraction expansion of the rational number p/q: $p/q = [a_0; a_1, \ldots, a_n]$. Let p_{n-1}/q_{n-1} and $p/q = p_n/q_n$ be the last two convergents of this expansion, where we assume that both α and p_{n-1}/q_{n-1} are simultaneously greater or smaller than the number p/q (if this does not happen then we replace the continued fraction with $p/q = [a_0; a_1, \ldots, a_n - 1, 1]$). Consider the number

$$\beta = \begin{pmatrix} p_n & p_{n-1} \\ q_n & q_{n-1} \end{pmatrix}^{-1} \alpha$$

$$= (-1)^{n-1} \begin{pmatrix} q_{n-1} & -p_{n-1} \\ -q_n & p_n \end{pmatrix} \alpha = -\frac{p_{n-1} - \alpha q_{n-1}}{p_n - \alpha q_n}, \qquad (2.19)$$

for which we have

$$\left| \beta + \frac{q_{n-1}}{q_n} \right| = \left| \frac{p_{n-1} - \alpha q_{n-1}}{\alpha q_n - p_n} + \frac{q_{n-1}}{q_n} \right|$$

$$= \frac{1}{q_n^2 |\alpha - p_n/q_n|} = \frac{1}{q^2 |\alpha - p/q|} > 2,$$

implying

$$|\beta| \geq \left| \beta + \frac{q_{n-1}}{q_n} \right| - \frac{q_{n-1}}{q_n} > 2 - \frac{q_{n-1}}{q_n} \geq 1. \qquad (2.20)$$

Comparing the latter inequality with (2.19) we deduce that

$$|p_{n-1} - \alpha q_{n-1}| > |p_n - \alpha q_n|;$$

hence

$$\left| \alpha - \frac{p_{n-1}}{q_{n-1}} \right| > \left| \alpha - \frac{p_n}{q_n} \right|.$$

However, the numbers α and p_{n-1}/q_{n-1} are both either greater or smaller than p/q, that is, α lies between p_{n-1}/q_{n-1} and $p/q = p_n/q_n$. But then $\beta > 0$ in accordance with (2.19), and so $\beta > 1$ by (2.20).

Let $[a_{n+1}; a_{n+2}, a_{n+3}, \dots]$ be the continued fraction of β; we have $a_{n+1} \geq 1$ in view of $\beta > 1$. Then

$$[a_0; a_1, \dots, a_n, a_{n+1}, \dots] = [a_0; a_1, \dots, a_n, \beta] = \begin{pmatrix} p_n & p_{n-1} \\ q_n & q_{n-1} \end{pmatrix} \beta = \alpha;$$

in other words, $p/q = p_n/q_n$ is indeed a convergent of α. □

2.7 Equivalent numbers

The set of matrices

$$\gamma = \begin{pmatrix} a & b \\ c & d \end{pmatrix}$$

with integer entries a, b, c, d and having determinant ± 1 (that is, $ad - bc = 1$ or -1) is a multiplicative group with identity (*neutral*) element

$$E = \begin{pmatrix} 1 & 0 \\ 0 & 1 \end{pmatrix}.$$

Indeed, the product of any two such matrices and the inverse of such a matrix again has integer entries and determinant equal to ± 1. This group is known as the *special linear group* (over the ring $\mathbb{Z}$) and is denoted by $SL_2(\mathbb{Z})$; in what follows we reserve the notation Γ for this group.

For an irrational number α, the action of an element $\gamma \in \Gamma$ is defined by the rule

$$\gamma\alpha = \frac{a\alpha + b}{c\alpha + d}. \tag{2.21}$$

EXERCISE 2.34 Show that the action is well defined, namely, that $E\alpha = \alpha$ and $\gamma(\delta\alpha) = (\gamma\delta)\alpha$ for all $\gamma, \delta \in \Gamma$.

We say that two irrational numbers α and β are *equivalent* if $\gamma\alpha = \beta$ for some $\gamma \in \Gamma$.

EXERCISE 2.35 Show that this relation is indeed an equivalence.

Example 2.36 For an irrational number α we have the representation

$$\alpha = [a_0; a_1, \dots, a_{n-1}, \alpha_n] = \frac{p_{n-1}\alpha_n + p_{n-2}}{q_{n-1}\alpha_n + q_{n-2}},$$

in accordance with Lemma 2.11. Define the $(n-1)th$ *continued transformation* of the number α by the equality

$$\gamma_{n-1} = \begin{pmatrix} p_{n-1} & p_{n-2} \\ q_{n-1} & q_{n-2} \end{pmatrix} = \prod_{j=0}^{n-1} \begin{pmatrix} a_j & 1 \\ 1 & 0 \end{pmatrix};$$

note that $\gamma_{n-1} \in \Gamma$ by Theorem 2.14. Then $\alpha = \gamma_{n-1}\alpha_n$, and hence α is equivalent to α_n for any $n \geq 1$. In other words, all complete quotients $\alpha_n, n = 1, 2, \ldots,$ are equivalent to each other.

The following theorem characterises the situation considered in Example 2.36.

Theorem 2.37   *Let $\alpha, \beta \in \mathbb{R} \setminus \mathbb{Q}$ and*

$$\alpha = \gamma\beta = \frac{a\beta + b}{c\beta + d} \qquad \text{for some} \quad \gamma = \begin{pmatrix} a & b \\ c & d \end{pmatrix} \in \Gamma.$$

Assume that $\beta > 1$ and $c > d > 0$. Then b/d and a/c are two consecutive convergents of α, say, p_{n-2}/q_{n-2} and p_{n-1}/q_{n-1}; furthermore, $\beta = \alpha_n$.

Proof   Note that a and c are relatively prime since $ad - bc = \pm 1$. Write a/c as the finite continued fraction

$$\frac{a}{c} = [a_0; a_1, \ldots, a_{n-1}] = \frac{p_{n-1}}{q_{n-1}}, \qquad a_{n-1} > 1,$$

where $a = p_{n-1}$ and $c = q_{n-1}$. Increasing by 1, if required, the length of the continued fraction for a/c (namely, replacing a_{n-1} with $a_{n-1} - 1, 1$), we obtain the equality

$$p_{n-1}q_{n-2} - q_{n-1}p_{n-2} = \epsilon, \tag{2.22}$$

where $\epsilon = ad - bc$. Then

$$ad - bc = p_{n-1}d - q_{n-1}b = \epsilon \tag{2.23}$$

and, comparing equations (2.22) and (2.23), we deduce that

$$p_{n-1}(d - q_{n-2}) = q_{n-1}(b - p_{n-2}). \tag{2.24}$$

Since p_{n-1} and q_{n-1} are coprime, we infer from (2.24) that q_{n-1} divides $d - q_{n-2}$; but $q_{n-2} \leq q_{n-1}$ and $0 < d < c = q_{n-1}$, that is, $|d - q_{n-2}| < q_{n-1}$, and hence $d - q_{n-2} = 0$. Then (2.24) implies that $b - p_{n-2} = 0$. Therefore

$$\alpha = \frac{a\beta + b}{c\beta + d} = \frac{p_{n-1}\beta + p_{n-2}}{q_{n-1}\beta + q_{n-2}} = [a_0; a_1, \ldots, a_{n-1}, \beta].$$

By the hypothesis $\beta > 1$, so the resulting expression is the continued fraction representing the number α and we have $\beta = \alpha_n$. In other words b/d and a/c are consecutive convergents of α.   □

Theorem 2.38 (Serret)   *Two numbers $\alpha, \beta \in \mathbb{R} \setminus \mathbb{Q}$ are equivalent if and only if there exist integers $n, m \geq 1$ such that $\alpha_n = \beta_m$. Equivalently, α and β are equivalent if and only if their continued fractions are*

$$\alpha = [a_0; a_1, a_2, \ldots] \qquad \text{and} \qquad \beta = [b_0; b_1, b_2, \ldots]$$

and $a_n = b_{n+l}$ for some $l \in \mathbb{Z}$ and all $n \geq N$.

Proof First assume that for some $n, m \geq 1$ we have $\alpha_n = \beta_m$, that is,

$$\alpha = [a_0; a_1, \dots, a_{n-1}, \alpha_n], \qquad \beta = [b_0; b_1, \dots, b_{m-1}, \beta_m],$$

and that $\alpha_n = \beta_m$. Since α is equivalent to α_n and β is equivalent to β_m (cf. Example 2.36), we conclude that α and β are equivalent.

Conversely, suppose that α and β are equivalent, that is,

$$\beta = \frac{a\alpha + b}{c\alpha + d} = \gamma\alpha, \qquad ad - bc = \pm 1.$$

Changing, if necessary, the signs of all entries of γ to their opposites, we may assume that $c\alpha + d > 0$. Let γ_{n-1} be the $(n-1)$th continued transformation of α; thus $\alpha = \gamma_{n-1}\alpha_n$. Then $\beta = \gamma\gamma_{n-1}\alpha_n$, and

$$\gamma\gamma_{n-1} = \begin{pmatrix} ap_{n-1} + bq_{n-1} & ap_{n-2} + bq_{n-2} \\ cp_{n-1} + dq_{n-1} & cp_{n-2} + dq_{n-2} \end{pmatrix} = \begin{pmatrix} a' & b' \\ c' & d' \end{pmatrix}.$$

We have

$$cp_{n-1} + dq_{n-1} = q_{n-1}\left(c\frac{p_{n-1}}{q_{n-1}} + d\right) = c',$$

$$cp_{n-2} + dq_{n-2} = q_{n-2}\left(c\frac{p_{n-2}}{q_{n-2}} + d\right) = d'.$$

(2.25)

Take n large enough that both p_{n-2}/q_{n-2} and p_{n-1}/q_{n-1} are close to α. Then $c' > 0$, $d' > 0$ and, in addition, $\alpha_n > 1$. Finally, Corollary 2.20 allows us to manipulate the parity of n in such a way that

$$c\frac{p_{n-2}}{q_{n-2}} < c\alpha < c\frac{p_{n-1}}{q_{n-1}};$$

then from (2.25) we have $c' > d'$ as $q_{n-2} < q_{n-1}$ (by Corollary 2.16). Thus, all the conditions of Theorem 2.37 are fulfilled, and we conclude that $\alpha_n = \beta_m$ for some m. This completes our proof of the theorem. □

EXERCISE 2.39 Let $\alpha = [a_0; a_1, a_2, a_3, \dots]$ be an irrational real number. Show the following:

(a) If $0 < \alpha < 1$ then $\alpha = [0; a_1, a_2, \dots]$ and $1/\alpha = [a_1; a_2, \dots]$.

(b) If $\alpha > 1$ then $\alpha = [a_0; a_1, \dots]$ and $1/\alpha = [0; a_0, a_1, \dots]$.

(c) If $-1/2 < \alpha < 0$ then $\alpha = [-1; 1, a_2, a_3, a_4]$ and $1/\alpha = [-(a_2 + 2); 1, a_3 - 1, a_4, \dots]$.
    (*Note*: this collapses to $[-(a_2 + 2); a_4 + 1, a_5, \dots]$ if $a_3 = 1$.)

(d) If $-1 < \alpha < -1/2$ then $\alpha = [-1; a_1, a_2, a_3, \dots]$, where $a_1 \geq 2$, and $1/\alpha = [-2; 1, a_1 - 2, a_2, \dots]$.
    (*Note*: this collapses to $[-2; a_2 + 1, a_3, \dots]$ if $a_1 = 2$.)

(e) If $\alpha < -1$ then $\alpha = [a_0; a_1, a_2, \dots]$, where $a_0 \le -2$ and $1/\alpha = [-1; 1, -(a_0 + 2), 1, a_1 - 1, a_2, \dots]$.
(*Note*: this collapses as follows: to $[-1; 2, a_1 - 1, a_2, \dots]$ if $a_0 = -2$ and $a_1 \ge 2$; to $[-1; 1, -(a_0 + 2), a_2 + 1, a_3, \dots]$ if $a_1 = 1$ and $a_0 \le -3$; and to $[-1; a_2 + 2, a_3, \dots]$ if $a_0 = -2$ and $a_1 = 1$.)

2.8 Continued fraction of a quadratic irrational

Let d be a positive integer. It can be seen that the set $\{x + y\sqrt{d} : x, y \in \mathbb{Q}\}$ forms a field. In what follows, we assume that this field does not coincide with $\mathbb{Q}$, in other words, that d is not a perfect square. Moreover, without loss of generality we may assume that the number d is square-free (that is, d is not divisible by a square > 1).

Note that 1 and $\sqrt{d}$ are linearly independent over $\mathbb{Q}$ (otherwise $\sqrt{d}$ would be rational). This implies that each element of the field possesses a *unique* representation in the form $x + y\sqrt{d}$ with $x, y \in \mathbb{Q}$. Let this field be denoted by $\mathbb{Q}(\sqrt{d})$ and define the *conjugate* of a number $\alpha = x + y\sqrt{d}$ to be $\overline{\alpha} = x - y\sqrt{d}$.

EXERCISE 2.40 Verify that

$$\overline{\alpha + \beta} = \overline{\alpha} + \overline{\beta} \quad \text{and} \quad \overline{\alpha\beta} = \overline{\alpha}\overline{\beta}.$$

Now define the *trace* and the *norm* of a number $\alpha \in \mathbb{Q}(\sqrt{d})$ by

$$\mathrm{Tr}(\alpha) = \alpha + \overline{\alpha} = 2x \in \mathbb{Q}, \qquad \mathrm{Norm}(\alpha) = \alpha\overline{\alpha} = x^2 - dy^2 \in \mathbb{Q}.$$

Then α and its conjugate $\overline{\alpha}$ are the roots of the quadratic polynomial

$$(x - \alpha)(x - \overline{\alpha}) = x^2 - \mathrm{Tr}(\alpha)\,x + \mathrm{Norm}(\alpha)$$

with rational coefficients; this characterises α as a *quadratic irrational*. Thus, a defining equation for the quadratic irrational α can be written in the form

$$\alpha^2 - 2x\alpha + (x^2 - dy^2) = 0;$$

taking $x^2 - dy^2 = c/a$ and $-2x = b/a$, where a, b, c are coprime integers and $a > 0$, we can represent the quadratic equation as

$$a\alpha^2 + b\alpha + c = 0$$

with coprime integers a, b, c, $a > 0$. Such a, b, c are determined by α uniquely. Finally, define the *discriminant* of a quadratic irrational α by the formula

$$D(\alpha) = b^2 - 4ac = 4a^2y^2d.$$

Since α is a real irrational number, we have $D(\alpha) > 0$.

We shall call α a *reduced* quadratic irrational if $\alpha > 1$ and $-1 < \overline{\alpha} < 0$ (equivalently, $-1/\overline{\alpha} > 1$).

EXERCISE 2.41   If α is a reduced quadratic irrational, show that $-1/\overline{\alpha}$ is reduced as well.

Theorem 2.42   *For a given positive integer D, there exist at most finitely many reduced elements of the field $\mathbb{Q}(\sqrt{d})$ whose discriminant is equal to D.*

Proof   Let α be a reduced number having discriminant $D(\alpha) = D$. Then

$$\alpha = \frac{-b + \epsilon\sqrt{D}}{2a} > 1 \qquad \text{and} \qquad -1 < \frac{-b - \epsilon\sqrt{D}}{2a} < 0, \qquad (2.26)$$

where $\epsilon = 1$ or -1. If $\epsilon = -1$ then we obtain $\alpha < 0$, which is impossible. Therefore $\epsilon = 1$ and, in accordance with $a > 0$ and (2.26),

$$b + \sqrt{D} < 2a < -b + \sqrt{D}. \qquad (2.27)$$

This means that $b < 0$; furthermore, the second inequality in (2.26) implies $-b < \sqrt{D}$. From these bounds on $|b|$ we conclude that there are finitely many possibilities for the quantity b to satisfy the inequalities (2.26). In turn, the inequality (2.27) retains only finitely many possibilities for the quantity $a > 0$ as well. Finally, the quantity $c \in \mathbb{Z}$ (if it exists) is subject to the relation $b^2 - 4ac = D$, and hence it is determined uniquely by the three quantities D, a and b. □

Lemma 2.43   *If α has discriminant $D > 0$ and β is equivalent to α then β has the same discriminant D.*

Proof   For $\alpha = x + y\sqrt{d}$ write

$$\alpha = \frac{A\beta + B}{E\beta + F}, \qquad AF - BE = \pm 1.$$

Then

$$a\left(\frac{A\beta + B}{E\beta + F}\right)^2 + b\frac{A\beta + B}{E\beta + F} + c = 0$$

is equivalent to the quadratic equation

$$a(A\beta + B)^2 + b(A\beta + B)(E\beta + F) + c(E\beta + F)^2$$
$$= (aA^2 + bAE + cE^2)\beta^2 + (2aAB + bAF + bBE + 2cEF)\beta$$
$$+ (aB^2 + bBF + cF^2) = 0$$

whose discriminant is equal to

$$(2aAB + bAF + bBE + 2cEF)^2$$
$$- 4(aA^2 + bAE + cE^2)(aB^2 + bBF + cF^2)$$
$$= b^2 - 4ac = D(\alpha),$$

and whose coefficients are coprime. (If there is a common multiple of the co-efficients then the inverse transformation

$$\beta = \frac{F\alpha - B}{-E\tau + A}$$

leads to the original quadratic equation for α, with coefficients a, b, c having the same common multiple.)   □

Theorem 2.44 *Let α be a real quadratic irrational number. Then*

  (i) *the number α_n, $n \geq 1$, in the continued fraction*

$$\alpha = [a_0; a_1, \ldots, a_{n-1}, \alpha_n]$$

   has the same discriminant as α;

 (ii) *if α is a reduced number then α_n is reduced for any $n \geq 1$ as well; and*

(iii) *if α is not necessarily reduced then α_n is reduced for all n sufficiently large.*

Proof Claim (i) follows from Lemma 2.43. Moreover, the defining procedure of the continued fraction for α implies $\alpha_n > 1$ for all $n \geq 1$.

  (ii) If α is reduced then $\alpha = a + 1/\beta$ for an integer $a \geq 1$ and a real $\beta > 1$; this implies $-1/\bar{\beta} = a - \bar{\alpha} > 1$, since $a \geq 1$ and $\bar{\alpha} < 0$. Therefore β is a reduced number as well.

  (iii) By Lemma 2.11,

$$\alpha = \frac{p_{n-1}\alpha_n + p_{n-2}}{q_{n-1}\alpha_n + q_{n-2}};$$

hence

$$\alpha_n = -\frac{q_{n-2}\alpha - p_{n-2}}{q_{n-1}\alpha - p_{n-1}}. \tag{2.28}$$

Therefore

$$\bar{\alpha}_n = -\frac{q_{n-2}\bar{\alpha} - p_{n-2}}{q_{n-1}\bar{\alpha} - p_{n-1}} = -\frac{q_{n-2}}{q_{n-1}} \frac{\bar{\alpha} - p_{n-2}/q_{n-2}}{\bar{\alpha} - p_{n-1}/q_{n-1}}. \tag{2.29}$$

Eventually the fractions p_{n-2}/q_{n-2} and p_{n-1}/q_{n-1} become close to α, so that both the numerator and denominator of the last fraction are close to $\overline{\alpha} - \alpha$; in particular, they are of the same sign. Consequently, $\overline{\alpha}_n < 0$. Furthermore,

$$\frac{\overline{\alpha} - p_{n-2}/q_{n-2}}{\overline{\alpha} - p_{n-1}/q_{n-1}} = 1 + \frac{p_{n-1}/q_{n-1} - p_{n-2}/q_{n-2}}{\overline{\alpha} - p_{n-1}/q_{n-1}}$$

$$= 1 + \frac{(-1)^n}{q_{n-1}q_{n-2}(\overline{\alpha} - p_{n-1}/q_{n-1})},$$

where we have used Corollary 2.15. Continuing (2.29) we find that

$$\overline{\alpha}_n + 1 = \frac{1}{q_{n-1}}\left(q_{n-1} - q_{n-2} + \frac{(-1)^n}{q_{n-1}(\overline{\alpha} - p_{n-1}/q_{n-1})}\right).$$

The expression

$$\frac{1}{q_{n-1}(\overline{\alpha} - p_{n-1}/q_{n-1})}$$

tends to 0 as $n \to \infty$, and hence its absolute value is less than 1 for all n sufficiently large. This implies $\overline{\alpha}_n + 1 > 0$ and demonstrates claim (iii). □

As a somewhat tangential application, we may iterate (2.28) to derive

Theorem 2.45 (Distance formula) *For $n \geq 0$ we have*

$$\alpha_1 \alpha_2 \cdots \alpha_n = \frac{(-1)^n}{p_{n-1} - q_{n-1}\alpha}, \tag{2.30}$$

with our previous conventions that $p_{-1} = 1$, $q_{-1} = 0$ and the empty product is interpreted as 1.

It turns out that one may usefully think of $\left|\log|p_{n-1} - q_{n-1}\alpha|\right|$ as measuring a weighted *distance* that the continued fraction has traversed in moving from α to α_n.

2.9 The Euler–Lagrange theorem

Let α be a real irrational number. We say that its continued fraction

$$[a_0; a_1, a_2, \dots]$$

is *periodic* if there exists an integer k such that $a_{n+k} = a_n$ for all n sufficiently large and *purely periodic* if $a_{n+k} = a_n$ for all $n \geq 0$; we call k the *primitive period* if it is the smallest positive integer with the above property.

The following standard notation is used for periodic continued fractions:

$$[a_0; a_1, \dots, a_r, \overline{a_{r+1}, \dots, a_{r+k}}],$$

where the vinculum (overbar) denotes the periodic repetition of the corresponding part. A continued fraction is purely periodic iff it can be written in the form $[\,\overline{a_0, a_1, \ldots, a_k}\,]$.

Lemma 2.46 *Let α be a reduced quadratic irrational and a an integer. Write $\alpha = a + 1/\beta$. Then β is reduced iff $a < \alpha < a + 1$, that is, iff $a = [\alpha]$.*

Proof If $a = [\alpha]$ then $\beta > 1$ and $-1/\overline{\beta} = a - \overline{\alpha} > a = [\alpha] \geq 1$, hence β is reduced.

Conversely, if $\alpha < a$ then $\beta < 0$, and if $a + 1 < \alpha$ then $\beta < 1$; thus β cannot be reduced if $a \neq [\alpha]$.   □

REMARK 2.47   We point out that the relation between α and β in Lemma 2.46 determines one of these numbers in terms of the other. Indeed,

$$-1/\overline{\beta} = a + \frac{1}{-1/\overline{\alpha}},$$

which implies that $a = [-1/\overline{\beta}]$. Moreover, $\overline{\alpha}$ (and hence α itself) is uniquely determined by $\overline{\beta}$ or, hence, by β.

We now come to a central result characterising quadratic irrationals in terms of the periodicity of their continued fractions. Recall, in contrast, that the eventual periodicity of its base-b expansion characterises the rationality of the number. This points to the power of continued fraction representations over b-ary ones.

Theorem 2.48 (Euler–Lagrange theorem)   *Let α be a real irrational number. The continued fraction for α is periodic iff α is a quadratic irrational. In the latter case, α is reduced iff its continued fraction is purely periodic.*

Proof   First assume that α is a quadratic irrational. By Theorem 2.44 the corresponding tails α_n are reduced for all $n \geq n_0$, while Theorem 2.42, together with Lemma 2.43, implies the finiteness of the reduced numbers that are equivalent to α. Therefore, for some $n \geq n_0$ and $k > 1$ we have $\alpha_n = \alpha_{n+k}$. This immediately implies the periodicity of the continued fraction. Furthermore, assume that α itself is reduced; by part (ii) of Theorem 2.44 all the α_n are reduced as well. As we already know, $\alpha_n = \alpha_{n+k}$ for some n and $k \geq 1$. From Lemma 2.46 and Remark 2.47 we conclude that α_{n-1} is uniquely determined by α_n and hence that $\alpha_{n-1} = \alpha_{n+k-1}$. Applying this descent n times, we finally arrive at $\alpha = \alpha_0 = \alpha_k$; in other words, the continued fraction is purely periodic.

Conversely, if a continued fraction is purely periodic then it may be written as

$$\alpha = [\,\overline{a_0; a_1, \ldots, a_k}\,] = [a_0; a_1, \ldots, a_k, \alpha].$$

The relation $\alpha = \gamma_k\alpha$ implies that α is a root of a quadratic equation with integer coefficients, while by claim (iii) of Theorem 2.44 the number $\gamma_k^n\alpha = \alpha$ is reduced. In the case of a periodic continued fraction, we write

$$\alpha = [a_0; a_1, \ldots, a_r, \overline{a_{r+1}, \ldots, a_{r+k}}] = [a_0; a_1, \ldots, a_r, \alpha_{r+1}],$$

where the purely periodic continued fraction $\alpha_{r+1} = [\,\overline{a_{r+1}, \ldots, a_{r+k}}\,]$ is, by the above argument, a (reduced) quadratic irrational. Since α and α_{r+1} are equivalent, the number α is a quadratic irrational as well. □

EXERCISE 2.49 Show that, if α is reduced and $\alpha = [\,\overline{a_0; a_1, \ldots, a_k}\,]$ then $-1/\overline{\alpha} = [\,\overline{a_k; a_{k-2}, \ldots, a_0}\,]$.

EXERCISE 2.50 ([125, Satz 3.9, p. 79]; see also Section 4.3 below) Let β be a real number. Show that β is the square root of a rational number > 1 iff there exist an integer $b_0 > 0$ and a finite (possibly empty) palindromic list of positive integers $b_1, \ldots, b_k$ such that $\beta = [b_0; \overline{b_1, \ldots, b_k, 2b_0}]$. (More about palindromes in continued fractions will be revealed in Section 2.14.)

Sketch of solution An equivalent way of saying that a list $b_1, b_2, \ldots, b_k$ is palindromic is that the matrix

$$\begin{pmatrix} a & b \\ c & d \end{pmatrix} = \begin{pmatrix} b_1 & 1 \\ 1 & 0 \end{pmatrix}\begin{pmatrix} b_2 & 1 \\ 1 & 0 \end{pmatrix} \cdots \begin{pmatrix} b_k & 1 \\ 1 & 0 \end{pmatrix}$$

is symmetric (that is, $b = c$). Writing

$$\beta = [b_0; \overline{b_1, \ldots, b_k, 2b_0}] = [b_0; b_1, \ldots, b_k, \beta + b_0]$$

$$= b_0 + \cfrac{1}{[b_1; b_2, \ldots, b_k, \beta + b_0]} = b_0 + \frac{c(\beta + b_0) + d}{a(\beta + b_0) + b}$$

we obtain a quadratic equation for β,

$$a\beta^2 + (b - c)\beta - b_0(ab_0 + b + c) = 0,$$

whose linear term vanishes iff $b = c$. □

2.10 Examples of non-periodic continued fractions

As we now know, continued fractions of quadratic irrationalities follow a simple pattern: they are periodic. Are there other examples of continued fractions that satisfy a clear law and correspond to 'meaningful' real numbers? This is in general an unresolved problem, but we will uncover some partial solutions; our immediate goal is to construct the continued fraction for e.

Consider the function

$$f(c, x) = 1 + \frac{1}{c}x + \frac{1}{c(c+1) \times 2!}x^2 + \cdots = \sum_{n=0}^{\infty} \frac{1}{c(c+1)\cdots(c+n-1)} \frac{x^n}{n!},$$

where we assume c to be an arbitrary real number different from $0, -1, -2, \ldots$ (to ensure that the series does not terminate). It is not hard to see that

$$f(c, x) = f(c+1, x) + \frac{x}{c(c+1)} f(c+2, x);$$

this can be written as

$$\frac{f(c+1, x)}{f(c, x)} = \frac{f(c+1, x)}{f(c+1, x) + \dfrac{x}{c(c+1)} f(c+2, x)}$$

$$= \frac{1}{1 + \dfrac{x}{c(c+1)} \dfrac{f(c+2, x)}{f(c+1, x)}}.$$

Now, this looks like a continued fraction expansion, but the term 1 in the denominator is not in the 'right' place. We set $x = z^2$ and transform the above relation further:

$$\frac{z}{c} \frac{f(c+1, z^2)}{f(c, z^2)} = \frac{1}{\dfrac{c}{z} + \dfrac{z}{c+1} \dfrac{f(c+2, z^2)}{f(c+1, z^2)}}.$$

This identity now has the right form to allow us to apply the general algebraic theory of continued fractions from Section 2.3.

By induction we get

$$\frac{z}{c} \frac{f(c+1, z^2)}{f(c, z^2)} = \left[0; \frac{c}{z}, \frac{c+1}{z}, \frac{c+2}{z}, \ldots, \frac{c+n}{z}, \alpha_{n+2} \right], \qquad (2.31)$$

where

$$\alpha_{n+2} = \frac{c+n+1}{z} \frac{f(c+n+1, z^2)}{f(c+n+2, z^2)}.$$

Thus, we do indeed get a continued fraction if we can specialise c and z in such a way that $\alpha_{n+2} \geq 1$ and all partial quotients $(c+n)/z$ are positive integers for each $n \geq 0$. These conditions are met for $c = 1/2$ and $z = 1/(2y)$, with $y \geq 1$ an arbitrary integer. The resulting continued fraction

$$\mathcal{L}(y) = [0; y, 3y, 5y, 7y, \ldots] \qquad (y = 1, 2, 3, \ldots)$$

is known as the *Lambert continued fraction* (which was actually known to Euler).

The quantity on the left-hand side of (2.31) admits a more familiar form. Indeed, for real w we have

$$\sinh w = \frac{e^w - e^{-w}}{2} = w\left(1 + \frac{w^2}{3!} + \frac{(w^2)^2}{5!} + \cdots\right)$$

$$= w \sum_{n=0}^{\infty} \frac{(w^2)^n}{(2n+1)!} = wf\left(\frac{3}{2}, \frac{w^2}{4}\right),$$

since the coefficient of $(w^2)^n$ in both power series is equal to

$$\frac{1}{\frac{3}{2}(\frac{3}{2}+1)\cdots(\frac{3}{2}+n-1)\times 4^n n!} = \frac{1}{3\times 5\cdots(2n+1)\times 2\times 4\cdots(2n)}$$

$$= \frac{1}{(2n+1)!}.$$

A similar argument leads to

$$\cosh w = \frac{e^w + e^{-w}}{2} = f\left(\frac{1}{2}, \frac{w^2}{4}\right).$$

Finally, letting $w = 1/y$, we arrive at the following conclusion.

Theorem 2.51 (Euler–Lambert)   *For every integer $y \geq 1$, we have*

$$\tanh\left(\frac{1}{y}\right) = \frac{e^{1/y} - e^{-1/y}}{e^{1/y} + e^{-1/y}} = [0; y, 3y, 5y, 7y, \dots];$$

in particular, for $y = 2$,

$$\frac{e-1}{e+1} = [0; 2, 6, 10, 14, \dots]. \tag{2.32}$$

The continued fraction for e itself is obtained in the next section and is derived from (2.32). Note that the numbers $(e-1)/(e+1)$ and e are *not* equivalent, since the determinant of the matrix $\left(\begin{smallmatrix} 1 & -1 \\ 1 & 1 \end{smallmatrix}\right)$ is 2.

The recursive method used in the proof of Theorem 2.51 can be applied for a more general class of functions – the so-called *hypergeometric functions*. However, just as in Theorem 2.51, one obtains continued fractions only for special cases of the parameters. We return to this in the final chapter, where we study the *Gauss continued fraction*.

EXERCISE 2.52 ([164]; see also the chapter notes below)   Find an expression for the value of the continued fraction

$$[0; y, 2y, 3y, 4y, \dots], \qquad y \in \mathbb{N}$$

in terms of *Bessel functions*.

2.11 The continued fraction for e

Following Euler we now prove the following statement.

Theorem 2.53   *If the continued fraction for e is given by*

$$e = [a_0; a_1, a_2, \ldots] = [2; 1, 2, 1, 1, 4, 1, 1, 6, 1, 1, 8, \ldots]$$

then $a_0 = 2$, $a_{3m-2} = a_{3m} = 1$ and $a_{3m-1} = 2m$ for $m \geq 1$.

Proof   Let r_n/s_n denote the nth convergent of the number

$$\alpha = \frac{e+1}{e-1}.$$

By Theorem 2.51, we have $\alpha^{-1} = [0; 2, 6, 10, 14, \ldots]$, and hence

$$\alpha = [2; 6, 10, 14, \ldots]$$

by Exercise 2.39. In addition,

$$e = \frac{\alpha + 1}{\alpha - 1}. \tag{2.33}$$

Let $\xi = [2; 1, 2, 1, 1, 4, 1, 1, 6, 1, \ldots]$, and let p_n/q_n denote the nth convergent of ξ. Let us show that, for $n \geq 0$, the following relation holds:

$$p_{3n+1} = r_n + s_n, \qquad q_{3n+1} = r_n - s_n. \tag{2.34}$$

These relations are easily checked for $n = 0, 1$. If $n \geq 2$ then

$$r_n = (4n+2)r_{n-1} + r_{n-2} \quad \text{and} \quad s_n = (4n+2)s_{n-1} + s_{n-2},$$

in accordance with Lemma 2.9. Now we multiply the recurrence relations for p_n (as well as for q_n) as follows,

$$
\begin{aligned}
p_{3n-3} &= p_{3n-4} + p_{3n-5} & &\text{by } 1, \\
p_{3n-2} &= p_{3n-3} + p_{3n-4} & &\text{by } -1, \\
p_{3n-1} &= 2np_{3n-2} + p_{3n-3} & &\text{by } 2, \\
p_{3n} &= p_{3n-1} + p_{3n-2} & &\text{by } 1, \\
p_{3n+1} &= p_{3n} + p_{3n-1} & &\text{by } 1,
\end{aligned}
$$

and sum them up. This gives us

$$p_{3n+1} = (4n+2)p_{3n-2} + p_{3n-5} \quad \text{and} \quad q_{3n+1} = (4n+2)q_{3n-2} + q_{3n-5}.$$

The equalities (2.34) are then deduced by induction on n. From (2.34) we conclude that

$$\frac{p_{3n+1}}{q_{3n+1}} = \frac{r_n + s_n}{r_n - s_n} = \frac{r_n/s_n + 1}{r_n/s_n - 1}. \tag{2.35}$$

Letting $n \to \infty$ in (2.35) we finally obtain

$$\xi = \frac{\alpha + 1}{\alpha - 1}, \tag{2.36}$$

and comparing (2.33) and (2.36) produces the required equality $\xi = e$.   □

As a corollary of Theorem 2.53 we deduce that e is neither rational nor a quadratic irrational.

EXERCISE 2.54 ([125, Bd. 1, Section 34])   Use the argument above to derive the *Perron continued fraction*,

$$e^{1/n} = [1; n - 1, 1, 1, 3n - 1, 1, 1, 1, 5n - 1, 1, 1, 1, 7n - 1, 1, 1, \ldots],$$

for integers $n \geq 1$. Deduce that e is not an nth root.

EXERCISE 2.55 ([143, p. 132])   Let $\alpha = 1/\sqrt{n}$ for a positive integer $n \geq 1$. Show that

$$\frac{\alpha}{\tanh \alpha} = [1; 3n, 5, 7n, 9, 11n, 13, \ldots].$$

Conclude that $e^{\sqrt{n}}$ is irrational for all integers $n \geq 1$.

EXERCISE 2.56   Show that

$$\tan\left(\frac{1}{n}\right) = [0; n - 1, 1, 3n - 2, 1, 5n - 2, 1, 7n - 2, 1, \ldots]$$

for integers $n \geq 1$.

EXERCISE 2.57 ([88])   Show that for integers $n \geq 0$ we have

$$e^{2/(2n+1)} = [1; n, 12n + 6, 5n + 2, 1, 1, 7n + 3, 36n + 18, 11n + 5, 1,$$

$$1, 13n + 6, 60n + 30, 17n + 8, 1, 1, \ldots]$$

$$= [(1, 3k(2n + 1) + n, 6(2k + 1)(2n + 1), (3k + 2)(2n + 1) + n, 1)_{k=0}^{\infty}].$$

2.12 The order of approximation of e by rationals

The aim of this section is to prove the following more refined result of C. Davis [54]; the use of the continued fraction for e from Theorem 2.53 is a crucial ingredient of the proof.

Theorem 2.58 (Davis' theorem)   *For any $\epsilon > 0$, there exist infinitely many rationals p/q that satisfy the inequality*

$$\left| e - \frac{p}{q} \right| < \left(\frac{1}{2} + \epsilon \right) \frac{\log \log q}{q^2 \log q};$$

conversely, the inequality

$$\left| e - \frac{p}{q} \right| < \left(\frac{1}{2} - \epsilon \right) \frac{\log \log q}{q^2 \log q}$$

has only finitely many solutions.

Theorem 2.58 means that the function

$$\phi(q) = \frac{\log \log q}{q^2 \log q}$$

is the best approximation order of e by rationals (recall the definition from Section 1.4).

Lemma 2.59   *For the continued fraction*

$$[a_0; a_1, a_2, \dots] = [2; 1, 2, 1, 1, 4, 1, 1, 6, 1, 1, 8, \dots]$$

for the number e, we have the asymptotic estimate

$$\log(a_1 a_2 \cdots a_n) = \frac{n \log n}{3} + O(n) \qquad as\ n \to \infty.$$

Proof   Set $k = \lfloor (n+1)/3 \rfloor$. We have

$$\log(a_1 a_2 \cdots a_n) = \sum_{l=1}^{k} \log(2l) = k \log 2 + \sum_{l=1}^{k} \log l.$$

To estimate the latter sum we use the inequalities

$$\int_{l-1}^{l} \log x \, dx < \int_{l-1}^{l} \log l \, dx = \log l = \int_{l}^{l+1} \log l \, dx < \int_{l}^{l+1} \log x \, dx,$$
$$(l = 1, \dots, k);$$

then

$$k \log k - k = \int_0^k \log x \, dx < \sum_{l=2}^{k} \log l$$
$$< \int_1^{k+1} \log x \, dx = (k+1)\log(k+1) - k. \quad (2.37)$$

On noting that $k = n/3 + O(1)$ we deduce the required asymptotics.   □

Proof of Theorem 2.58   Let us start with the second part of the theorem. If a rational number p/q is not a convergent of e then, by Lemma 2.33, we have

$$\left| e - \frac{p}{q} \right| \geq \frac{1}{2q^2}$$

and, in particular, the inequality

$$\left| e - \frac{p}{q} \right| < \left(\frac{1}{2} - \epsilon \right) \frac{\log \log q}{q^2 \log q} \tag{2.38}$$

does not have solutions in such rationals. In the case $p/q = p_n/q_n$ with either $n = 3m - 3$ or $n = 3m - 1$ we have $a_{n+1} = 1$; hence

$$\left| e - \frac{p}{q} \right| > \frac{1}{3q^2}$$

by Lemma 2.30, which contradicts (2.38) for $q > 2$. Finally, if $p/q = p_n/q_n$ with $n = 3m - 2$ then $a_{n+1} = a_{3m-1} = 2m$, implying that

$$\left| e - \frac{p}{q} \right| = \left| e - \frac{p_{3m-2}}{q_{3m-2}} \right| > \frac{1}{(2m + 2)q_{3m-2}^2}. \tag{2.39}$$

Using the estimates in (2.18), $\log F_n = O(n)$ and Lemma 2.59, we deduce that

$$\log q_{3m-2} = \log q_n = \log(a_1 \cdots a_n) + O(n)$$
$$= \frac{n \log n}{3} + O(n) = m \log m + O(m).$$

Hence

$$\frac{\log q_{3m-2}}{\log \log q_{3m-2}} \sim m \qquad \text{as } m \to \infty,$$

and substitution of this asymptotic estimate into (2.39) gives

$$\left| e - \frac{p}{q} \right| > \frac{\log \log q}{2q^2 \log q}(1 + o(1)).$$

This demonstrates the second part of the theorem.

As for the first part, note that choosing $p/q = p_{3m-2}/q_{3m-2}$ in Lemma 2.30 (the estimate in (2.17) from above) and using Lemma 2.59 gives us

$$\left| e - \frac{p}{q} \right| < \frac{\log \log q}{2q^2 \log q}(1 + o(1)).$$

This completes the proof of the theorem.   □

EXERCISE 2.60 (Estimation of $e^{1/y}$)   Using Perron's continued fraction for $e^{1/y}$ from Exercise 2.54, extend Theorem 2.58 to rational approximations to $e^{1/y}$ with $y \in \mathbb{N}$.

2.13 Bounded partial quotients

Let α be a real irrational number with continued fraction $[a_0; a_1, a_2, \ldots]$. If

$$K(\alpha) = \sup_{k \geq 1} a_k$$

is finite, we say that α has *bounded partial quotients* (or is *of finite type* [94] or is *badly approximable* [149]). Otherwise, if $K(\alpha) = \infty$ then we say that α has *unbounded partial quotients*. We define $\mathcal{B}$ to be the set of all real numbers with bounded partial quotients. As we will see in Remark 3.2, it is an uncountable set of Lebesgue measure zero.

Not many 'naturally occurring' irrational numbers are known to be in $\mathcal{B}$ other than the quadratic irrationals. We will see some additional examples in Chapter 6. The status of the algebraic numbers of degree > 2 is currently unresolved. It is conjectured that all these numbers have unbounded partial quotients.

In this section we prove some properties of the numbers with bounded partial quotients. In particular, we show that $\mathcal{B}$ is closed under the *linear fractional transformations*, that is, the maps $\alpha \mapsto (a\alpha + b)/(c\alpha + d)$ for integers a, b, c, d with $ad - bc \neq 0$.

Recall that if $\alpha = [a_0; a_1, a_2, \ldots]$ then α_n denotes the nth complete quotient $[a_n; a_{n+1}, a_{n+2}, \ldots]$. We define the *norm*

$$\|\alpha\| = \min(\alpha - \lfloor \alpha \rfloor, \lceil \alpha \rceil - \alpha) = \min(\{\alpha\}, 1 - \{\alpha\}),$$

that is, the distance from α to the nearest integer.

Lemma 2.61 *Suppose that $q\|q\alpha\| \geq 1/r$ for some $r \geq 1$ and all $q \geq 1$. Then $K(\alpha) < r$.*

Proof Let p_n/q_n be a convergent to α with $n \geq 1$. Since

$$\left| \frac{p_n}{q_n} - \alpha \right| < \frac{1}{a_{n+1} q_n^2}$$

from Theorem 2.25, we see that $q_n|p_n - q_n\alpha| < 1/a_{n+1}$. Now, clearly $\|q_n\alpha\| \leq |p_n - q_n\alpha|$, so that

$$\frac{1}{r} \leq q_n\|q_n\alpha\| \leq q_n|p_n - q_n\alpha| < \frac{1}{a_{n+1}};$$

hence $a_{n+1} < r$ for all $n \geq 0$ and the estimate $K(\alpha) < r$ follows. □

Lemma 2.62 *Suppose that $K(\alpha) \leq r$ for some $r \geq 1$. Then for all $q \geq 1$ we have $q\|q\alpha\| \geq 1/(r + 2)$.*

Proof We will proceed by contradiction. Suppose there exists a $q \geq 1$ such that $q\|q\alpha\| < 1/(r+2)$. Let p be an integer such that $\|q\alpha\| = |q\alpha - p|$. Then $q|q\alpha - p| < 1/(r+2)$, implying that

$$\left|\alpha - \frac{p}{q}\right| < \frac{1}{(r+2)q^2} < \frac{1}{2q^2},$$

from which it follows by Lemma 2.33 that p/q is a convergent to α, say $p/q = p_n/q_n$. Thus, $p = ap_n$ and $q = aq_n$ for some integer $a \geq 1$.

Now, from Lemma 2.24 we have

$$|q_n\alpha - p_n| = \frac{1}{\alpha_{n+1}q_n + q_{n-1}};$$

hence

$$\frac{1}{r+2} > q\|q\alpha\| = q|q\alpha - p| = aq_n|aq_n\alpha - ap_n|$$

$$\geq q_n|q_n\alpha - p_n| \geq \frac{1}{\alpha_{n+1} + q_{n-1}/q_n} \geq \frac{1}{(a_{n+1}+1)+1}.$$

Thus $a_{n+1} > r$ and so $K(\alpha) > r$. $\square$

We say α is of *type $< r$* if $q\|q\alpha\| \geq 1/r$ for all integers $q \geq 1$.

Lemma 2.63 *Let a and b be integers with $a \geq 1$, $|b| \geq 1$. If α is of type $< M$ then $a\alpha/b$ is of type $< |ab|M$.*

Proof Again we proceed by contradiction. Assume that

$$q\left\|q\frac{a}{b}\alpha\right\| < \frac{1}{|ab|M}$$

for some $q \geq 1$.

Now there exists an integer p such that

$$\left\|q\frac{a}{b}\alpha\right\| = \left|q\frac{a}{b}\alpha - p\right|.$$

Thus,

$$\left|q\frac{a}{b}\alpha - p\right| < \frac{1}{|ab|M}$$

and so, multiplying by $|ab|$, we get $qa|qa\alpha - pb| < 1/M$. Hence we deduce that $qa\|qa\alpha\| < 1/M$. $\square$

Corollary 2.64 *For integers a, b with $b > 0$ we have*

$$K\left(\frac{a}{b}\alpha\right) < |ab|\,(K(\alpha)+2).$$

Lemma 2.65 *Let a and b be integers with* $|a| \geq 1$, $b \geq 1$. *If* α *is of type* $< M$ *then* $\alpha + a/b$ *is of type* $< b^2 M$.

Proof Once again we use contradiction. Assume that

$$q \left\| q\left(\alpha + \frac{a}{b}\right) \right\| < \frac{1}{b^2 M}$$

for some $q \geq 1$.

There exists an integer p such that $\|q(\alpha + a/b)\| = |q(\alpha + a/b) - p|$; thus,

$$q \left| q\left(\alpha + \frac{a}{b}\right) - p \right| < \frac{1}{b^2 M}$$

and so, multiplying by b^2, we get $qb|qb\alpha + qa - pb| < 1/M$. Hence we do indeed have $qb\|qb\alpha\| < 1/M$. □

Corollary 2.66 *For integers* a, b *with* $b > 0$ *we have*

$$K\left(\alpha + \frac{a}{b}\right) < b^2(K(\alpha) + 2).$$

Lemma 2.67 *The following estimates hold:*

$$K\left(\frac{1}{\alpha}\right) \leq \begin{cases} K(\alpha) & \text{if } 0 < \alpha < 1; \\ \max(K(\alpha), \lfloor \alpha \rfloor) & \text{if } \alpha > 1; \\ K(\alpha) + 1 & \text{if } -1 < \alpha < 0; \\ \max(K(\alpha) + 2, -\lfloor \alpha \rfloor - 2) & \text{if } \alpha < -1. \end{cases}$$

Proof This follows immediately from Exercise 2.39. □

Theorem 2.68 *Let* a, b, c, d *be integers with* $ad - bc \neq 0$. *Then* α *has bounded partial quotients iff* $(a\alpha + b)/(c\alpha + d)$ *has bounded partial quotients.*

Proof Sufficiency follows from Corollaries 2.64 and 2.66, where for $c \neq 0$ we write

$$\beta = \frac{a\alpha + b}{c\alpha + d} = \frac{b - (ad/c)}{c\alpha + d} + \frac{a}{c}.$$

Proving necessity requires us to write $\alpha = (d\beta - b)/(-c\beta + a)$ and apply the argument above with β replaced by α. □

2.14 In the footsteps of Maillet

It is easy to give examples of irrational real numbers where both x and x^2 have bounded partial quotients, such as $x = (1 + \sqrt{5})/2$. But much more is true. A theorem of Schmidt [148] implies that there exist uncountably many real numbers x for which $x, x^2, x^3, \ldots$ all have bounded partial quotients (for a weaker result, see [53]). There are also numbers known for which both x and x^2 have unbounded partial quotients; one example, as we have seen in Theorem 2.53 and Exercise 2.57, is $x = e$. Moreover, if x is a Liouville number, such as $\sum_{n=0}^{\infty} 2^{-n!}$ (see Exercise 1.40), then it is easy to see that every positive power of x has unbounded partial quotients.

In this section we shall show how to construct real numbers x with bounded partial quotients such that x^2 has unbounded partial quotients. We also show how to construct a number x with bounded partial quotients such that *every* nonzero even power of x has unbounded partial quotients.

The basic idea of the construction can be found in some little known work of Edmond Théodore Maillet (1865–1938) (see, for example, [125, Section IV.36] and [109]); it goes as follows. We construct a sequence of distinct rational numbers $x_1, x_2, x_3, \ldots$ and a sequence of distinct quadratic irrationals $y_1, y_2, y_3, \ldots$ such that

$$\lim_{i \to \infty} x_i = \lim_{i \to \infty} y_i$$

and for all $i \geq 1$ the following hold:

1. x_i has a simple continued fraction of the form $[1; b_i]$, where b_i is a finite list of 1s and 2s that starts with a 1 and ends with a 2;
2. the continued fraction for x_i is a *prefix* of that for x_{i+1};
3. y_i^2 is a rational number;
4. y_{i+1}^2 is an excellent approximation to y_i^2 – so good, in fact, that the continued fractions of y_{i+1}^2 and y_i^2 coincide at least as far as the penultimate partial quotient in the expansion for y_i^2; if the expansions first differ at partial quotient a_j then either the $(j+1)$th or $(j+2)$th partial quotient of y_{i+1}^2 is very large.

Our construction uses strings (or lists) of partial quotients. If b is a string then by ${}^t b$ we mean the reverse (or transpose) of b; for example, ${}^t(1, 2, 3) = (3, 2, 1)$. If $b = {}^t b$ then we say that b is a *palindrome* (see Exercise 2.50). We define $|b|$ to be the number of terms in the string b. By $\{b\}^k$ we mean the string in which b is repeated k times (cf. Exercise 2.10), so that $|\{b\}^k| = k|b|$; by $\{b\}^\infty$ we denote the string $b, b, b, \ldots$, which is infinite if $|b| > 0$. Reprising the definitions in Section 2.9, if an infinite string c can be written in the form $\{b\}^\infty$ for some finite nonempty string b then we say that c is *purely periodic*. The *period* of a purely periodic string c is the length of the shortest string b such that $c = \{b\}^\infty$.

Here are more details. Set $b_1 = (1, 2)$. Now, for $i = 1, 2, 3, \ldots$, define the following: $m_i = |b_i|$, $x_i = [1; b_i]$, $y_i = [1; \{b_i\}^\infty] = [1; b_i, b_i, b_i, \ldots]$ and $r_i =$

$y_i^2 = e_i/f_i$, where $\gcd(e_i, f_i) = 1$. Choose an integer $n_i \geq 1$ such that if $g_i/h_i = [1; \{b_i\}^{n_i}]$ then $h_i^2 > 4(100^i + 2)f_i^2$. Finally, let $b_{i+1} = (\{b_i\}^{n_i}, \{2\}^{m_i}, \{^t b_i\}^{n_i}, 2)$.

We then have the following theorem.

Theorem 2.69 (Bounded quotients with unbounded squares) *Let x_i, r_i be defined as in the preceding construction, and define*

$$x = \lim_{i \to \infty} x_i \quad and \quad r = \lim_{i \to \infty} r_i.$$

Then the partial quotients of x are all 1s and 2s, the partial quotients of r are unbounded and $r = x^2$.

To prove the correctness of the construction of the theorem, we need some lemmas.

Let $r = [a_0; a_1, \ldots, a_k] \neq 0$ be the continued fraction of a rational number, with $a_k \geq 2$ if $k \geq 1$. Then define length(r) = $k + 1$.

Lemma 2.70 *Suppose that r is a rational number and that $[a_0; a_1, \ldots, a_k]$ is its continued fraction, where the last partial quotient $a_k \geq 2$ (if $k \neq 0$). Let p_k/q_k be the kth convergent to r. Let t be an integer ≥ 2, and suppose that s is a real number, with*

$$0 < |r - s| < \frac{1}{(t+1)q_k^2}.$$

Then

(i) *if $r < s$ and k is even, or if $r > s$ and k is odd, we have $s = [b_0; b_1, b_2, \ldots]$, where $b_i = a_i$ for $0 \leq i \leq k$ and $b_{k+1} \geq t$;*

(ii) *if $r < s$ and k is odd, or if $r > s$ and k is even, we have $s = [b_0; b_1, b_2, \ldots]$, where $b_i = a_i$ for $0 \leq i \leq k - 1$, $b_k = a_k - 1$, $b_{k+1} = 1$ and $b_{k+2} \geq t$.*

Furthermore, if s is rational then length(r) < length(s).

Proof We will treat only the case $r < s$: the case $r > s$ is similar and is left to the reader.

(i) Assume k is even. Then we have that the first $k + 2$ terms of the continued fraction expansion for s are $a_0, a_1, \ldots, a_k, b$ iff

$$[a_0; a_1, \ldots, a_k, b, 1] \leq s \leq [a_0; a_1, \ldots, a_k, b].$$

Hence the first $k + 2$ terms of the continued fraction for s are of the form $a_0, a_1, \ldots, a_k, b$, with $b \geq t$, iff

$$[a_0; a_1, \ldots, a_k] \leq s \leq [a_0; a_1, \ldots, a_k, t],$$

iff

$$r = \frac{p_k}{q_k} \leq s \leq \frac{tp_k + p_{k-1}}{tq_k + q_{k-1}}$$

and iff

$$0 \leq s - r \leq \frac{tp_k + p_{k-1}}{tq_k + q_{k-1}} - \frac{p_k}{q_k} = \frac{1}{q_k(tq_k + q_{k-1})},$$

where we have used Theorem 2.14. Hence, if

$$0 < s - r < \frac{1}{(t+1)q_k^2},$$

it follows that the continued fraction for s begins with the partial quotients $a_0, a_1, \ldots, a_k, b$, with $b \geq t$.

(ii) Now assume that k is odd. Then the first $k + 3$ terms of the continued fraction expansion for s are $a_0, a_1, \ldots, a_{k-1}, a_k - 1, 1, b$ iff

$$[a_0; a_1, \ldots, a_{k-1}, a_k - 1, 1, b, 1] \leq s \leq [a_0; a_1, \ldots, a_{k-1}, a_k - 1, 1, b].$$

Hence the first $k + 3$ terms of the continued fraction for s are $a_0, a_1, \ldots, a_{k-1}$, $a_k - 1, 1, b$, with $b \geq t$, iff

$$[a_0; a_1, \ldots, a_{k-1}, a_k - 1, 1] \leq s \leq [a_0; a_1, \ldots, a_{k-1}, a_k - 1, 1, t],$$

iff

$$r = \frac{p_k}{q_k} \leq s \leq \frac{(t+1)p_k - p_{k-1}}{(t+1)q_k - q_{k-1}}$$

and iff

$$0 \leq s - r \leq \frac{(t+1)p_k - p_{k-1}}{(t+1)q_k - q_{k-1}} = \frac{1}{q_k((t+1)q_k - q_{k-1})},$$

where again we have used Theorem 2.14. Hence, if

$$0 < s - r < \frac{1}{(t+1)q_k^2},$$

it follows that the continued fraction for s begins with the partial quotients $a_0, a_1, \ldots, a_{k-1}, a_k - 1, 1, b$, with $b \geq t$. □

Lemma 2.71 *Let α and β be two positive real numbers such that the first $k+1$ terms of their continued fractions coincide and are equal to $a_0, a_1, \ldots, a_k$. Let $p_k/q_k = [a_0; a_1, \ldots, a_k]$. Then $|\alpha - \beta| \leq 1/q_k^2$.*

Proof Assume that k is odd. (The case where k is even is similar.) Then

$$[a_0; a_1, \ldots, a_k, 1] \leq \alpha, \beta \leq [a_0, a_1, \ldots, a_k].$$

58   *Continued fractions*

Hence we have

$$\frac{p_k + p_{k-1}}{q_k + q_{k-1}} \le \alpha, \beta \le \frac{p_k}{q_k}.$$

It follows that

$$|\alpha - \beta| \le \left|\frac{p_k + p_{k-1}}{q_k + q_{k-1}} - \frac{p_k}{q_k}\right| \le \frac{1}{q_k^2},$$

where we have used Theorem 2.14.   □

We now prove the correctness of the construction.

Proof of Theorem 2.69 First, we claim the following are true for all $i \ge 1$:

(i) b_i is a string of 1s and 2s which ends in a 2;
(ii) b_i is a palindrome if the last symbol is removed;
(iii) b_i is a prefix of b_{i+1};
(iv) $m_i < m_{i+1}$;
(v) $1 < x_i, y_i < 2$;
(vi) r_i is a rational number.

Parts (i)–(v) follow easily from the definition of b_i; for part (vi), combine (ii) with Exercise 2.50.

Next, we argue that all the y_i are distinct, that is, if $i \ne j$ then $y_i \ne y_j$. For this, it suffices to show that $\{b_i\}^\infty \ne \{b_j\}^\infty$. We do this by showing that the period p_i of $\{b_i\}^\infty$ satisfies $m_{i-1} < p_i \le m_i$ for $i \ge 2$. The upper bound is clear, since $m_i = |b_i|$.

For the lower bound, observe that by construction b_i starts with a 1 but contains a block of m_{i-1} consecutive 2s. It follows that p_i is at least $m_{i-1} + 1$. Hence $p_1 < p_2 < p_3 < \cdots$, and so the y_i are all distinct.

It is clear that

$$x = \lim_{i\to\infty} x_i \quad \text{and} \quad y = \lim_{i\to\infty} y_i.$$

exist. By construction, the continued fractions for y_i and x_{i+1} agree on partial quotients a_0 through $a_{n_i m_i}$, and so by Lemma 2.71 we have

$$|y_i - x_{i+1}| \le \frac{1}{h_i^2} < 100^{-i}.$$

It follows that $x = y$, and so if $r = \lim_{i\to\infty} r_i$ then $r = x^2$.

Finally, it remains to see whether r has unbounded partial quotients. We apply Lemma 2.70 with $r = r_i$ and $s = r_{i+1}$. Since the y_i are all distinct, we know that $r \ne s$ and, by result (v) above, r is a rational number. By construction,

$$|y_i - y_{i+1}| \le \frac{1}{h_i^2} < \frac{1}{4(100^i + 2)f_i^2}.$$

Hence, using result (iv) above, we get

$$|r - s| = |r_i - r_{i+1}| = |y_i^2 - y_{i+1}^2|$$

$$= (y_i + y_{i+1})|y_i - y_{i+1}| < 4|y_i - y_{i+1}| < \frac{1}{(100^i + 2)f_i^2}.$$

By Lemma 2.70, it now follows that the continued fractions for r_i and r_{i+1} coincide up to the penultimate partial quotient of r_i, and the continued fraction for r_{i+1} contains a partial quotient b of size $\geq 100^i + 1$. Furthermore, length(r_{i+2}) > length(r_{i+1}) > length(r_i), so this new large partial quotient b introduced in the continued fraction for r_{i+1} can be reduced by at most 1 in the continued fraction for r_{i+2} (and this reduction can occur only if b is the last partial quotient in the continued fraction r_{i+1}). The partial quotient b cannot be reduced further in $r_{i+3}, r_{i+4}, \ldots$ It follows that the continued fraction for r_{i+2} contains a partial quotient $\geq 100^i$ and that this partial quotient also appears in r. Hence r has unbounded partial quotients. $\quad\square$

REMARK 2.72 It seems worthwhile pointing out which parts of the above construction are crucial and which parts are somewhat arbitrary. The definition of b_1 is more or less arbitrary as long as it begins with 1, ends with 2 and is a palindrome when the last term is deleted. Similarly, the choice of the term 100^i in the denominator is more or less arbitrary; any function tending to infinity sufficiently quickly could be substituted.

Example 2.73 Let us look at an example of the construction. We have

---

$b_1 = (1, 2)$;
$x_1 = [1; b_1] = [1; 1, 2] = 5/3$;
$y_1 = [1; b_1, b_1, b_1, \ldots] = \sqrt{3}$;
$r_1 = 3 = 3/1 = [3]$;
$m_1 = 2$;
$n_1 = 3$;

$b_2 = (b_1, b_1, b_1, 2, 2, {}^t b_1, {}^t b_1, {}^t b_1, 2) = (1, 2, 1, 2, 1, 2, 2, 2, 2, 1, 2, 1, 2, 1, 2)$;
$x_2 = 52\,472/30\,297$;
$y_2 = \sqrt{6\,653/2\,218}$;
$r_2 = 6\,653/2\,218 = [2; 1, 2\,217]$;
$m_2 = 15$;
$n_2 = 2$;

$b_3 = (b_2, b_2, \{2\}^{15}, {}^t b_2, {}^t b_2, 2);$

$x_3 = 400\,183\,051\,571\,243\,862\,647\,721/2\,310\,631\,559\,280\,479\,904\,375\,689;$

$y_3 = \sqrt{70\,471\,919\,435\,937\,751\,341\,509/23\,494\,170\,646\,161\,120\,172\,022};$

$r_3 = [2; 1, 2\,216, 1, 365\,316\,604\,851, 1, 1, 2, 3, 5, 4, 1, 3, 3, 2, 1, 136, 1, 2, 1, 2];$

$m_3 = 76;$

$n_3 = 2.$

Continuing in this fashion, we construct a number $x = \lim_{i \to \infty} x_i$, given by

$$x = [1; 1, 2, 1, 2, 1, 2, 2, 2, 2, 1, 2, 1, 2, 1, 2, 1, 2, 1, 2,$$
$$1, 2, 2, 2, 2, 1, 2, 1, 2, 1, 2, 2, 2, 2, 2, 2, 2, 2, 2, \ldots]$$

and for which

$$x^2 = [2; 1, 2\,216, 1, 365\,316\,604\,851, 1, 1, 2, 3, 5, 4, 1, 3, 3, 2, 1, 136,$$
$$1, 2, 1, 1, 1, N, \ldots],$$

where

$N = 110\,153\,056\,861\,410\,353\,112\,260\,022\,760\,667\,309\,348\,898\,319\,235\,606\,578.$

1. By choosing a more rapidly growing function one can easily ensure that the resulting value of r is *transcendental* (not a root of any algebraic equation with rational coefficients).
2. By choosing to introduce either a block of 2s or 3s instead of a block of 2s in the middle at each step, the construction can be easily modified to give *uncountably* many numbers x with the desired properties.
3. By replacing y_i^2 with y_i^d in the construction given at the start of this section, and suitably modifying the inequality test, we can generate examples of numbers x with bounded partial quotients for which x^d has unbounded partial quotients for any even positive integer d.

In fact, we can generate numbers x with bounded partial quotients for which all nonzero even powers of x have unbounded partial quotients. To do this, we modify the construction as follows.

Set $b_1 = (1, 2)$. Now, for $i = 1, 2, 3, \ldots$, define the following: $m_i = |b_i|$, $x_i = [1; b_i]$ and $y_i = [1; \{b_i\}^\infty] = [1; b_i, b_i, b_i, \ldots]$. For $j = 1, 2, \ldots, i$ define $r_{i,j} = e_{i,j}/f_{i,j} = y_i^{2j}$. Choose an integer $n_i \geq 1$ such that if $g_i/h_i = [1; \{b_i\}^{n_i}]$ then

$$h_i^2 > \max_{1 \leq j \leq i} j\,2^{2j}(100^i + 2)f_{i,j}^2.$$

Finally, let $b_{i+1} = (\{b_i\}^{n_i}, \{2\}^{m_i}, \{{}^t b_i\}^{n_i}, 2).$

Theorem 2.74 (Unbounded even powers) *If x_i is defined as given in the preceding construction, and $x = \lim_{i\to\infty} x_i$, then x has partial quotients consisting of 1s and 2s, but every nonzero even power of x has unbounded partial quotients.*

Proof The proof is largely the same as the proof of Theorem 2.69 and is left to the reader. □

Even more is possible. By altering the construction of the previous theorem to utilise a list of *all* rational functions with integer coefficients, not just powers, we can prove the existence of *a real number x with bounded partial quotients such that every rational function of x^2 has unbounded partial quotients.*

Notes

The 2×2 matrix approach to continued fractions was originally developed by Hurwitz before 1916 (matrices themselves were then quite new); see [76]. It was later rediscovered by Frame [63] and Kolden [84], independently.

Smith's representation of the continuants given in Exercise 2.23 allowed him to produce an elegant and constructive proof of Fermat's 'two squares' theorem: *every prime p congruent to 1 (mod 4) can be expressed as a sum of two squares.* We refer the reader to [46] for the details of this derivation and some historical remarks.

The correspondence between continued fractions and matrices discussed in Section 2.7 in effect identifies *all nonsingular 2×2 matrices*

$$\gamma = \begin{pmatrix} a & b \\ c & d \end{pmatrix}$$

(not necessarily from $\Gamma = \mathrm{SL}_2(\mathbb{Z})$) with linear fractional transformations (2.21), because $k\gamma$ for an arbitrary scalar $k \neq 0$ acts on a real number α independently of k.

Then *any* sequence

$$\gamma_n = \begin{pmatrix} a_n & b_n \\ c_n & d_n \end{pmatrix}$$

of nonsingular 2×2 matrices *such that a_n/c_n and b_n/d_n have a common limit as $n \to \infty$* yields a continued fraction expansion. For example, if

$$\begin{pmatrix} a_n & b_n \\ c_n & d_n \end{pmatrix} = \prod_{m=0}^{n} \begin{pmatrix} 2m+1+z & 2m+1 \\ 2m+1 & 2m+1-z \end{pmatrix}$$

then $a_n d_n - b_n c_n = (-1)^{n+1} z^{2(n+1)}$ shows that the z-formal power series a_n/c_n and b_n/d_n coincide in the limit. Here $a_n(z) = d_n(-z)$ and $b_n(z) = c_n(-z)$ and we need to confirm only that both $a_n(z)$ and $b_n(z)$ times $e^{-z/2} n!/(2n+1)!$ converge to 1 as $n \to \infty$; so here the common limit is e^z.

Since

$$\begin{pmatrix} 2m+2 & 2m+1 \\ 2m+1 & 2m \end{pmatrix} = \begin{pmatrix} 1 & 1 \\ 1 & 0 \end{pmatrix} \begin{pmatrix} 2m & 1 \\ 1 & 0 \end{pmatrix} \begin{pmatrix} 1 & 1 \\ 1 & 0 \end{pmatrix}$$

we obtain the main result of Section 2.11:

$$e - 1 = [1; 1, 2, 1, 1, 4, 1, 1, 6, 1, 1, 8, \ldots] = [1; \overline{1, 2n, 1}]_{n=1}^{\infty}.$$

Another recent self-contained proof by Cohn of the continued fraction for e is to be found in [47].

A lovely classical result [125], re-observed by Schroeppel in 1972, is that for real a and $b \neq 0$ any arithmetic simple continued fraction is expressible as a ratio of modified Bessel functions of the second kind:

$$[a + b, a + 2b, a + 3b, \ldots] = \frac{I_{a/b}(2/b)}{I_{1+a/b}(2/b)}.$$

In particular,

$$[1, 2, 3, \ldots] = \frac{I_0(2)}{I_1(2)}.$$

In our derivation of Davis' Theorem 2.58 we followed Tasoev [159], who proves a variant of the result for more general pseudo-periodic continued fractions. In a related way, [28, Theorem 11.1] yields integers

$$p_n = \sum_{k=0}^{n} \frac{(2n-k)!}{(n-k)!k!}, \qquad q_n = \sum_{k=0}^{n} (-1)^k \frac{(2n-k)!}{(n-k)!k!}, \qquad (2.40)$$

such that

$$\left| e - \frac{p_n}{q_n} \right| = \frac{\log \log q_n}{2q_n^2 \log q_n} (1 + o(1)).$$

The majority of other mathematical constants, including π, have no known simple patterns in their continued fraction expansions. However, suitable generalizations of continued fractions, such as the irregular continued fractions we consider in Chapter 9, do sometimes exhibit simple patterns; one example of such a continued fraction is given in (9.16).

For more results regarding bounded partial quotients, one may consult the survey [151]. Exercise 2.39 was based on [83, Exercise 4.5.3.10]. A different proof of Theorem 2.68 is given in [92]. Some results in Section 2.14 can be found in [37].

A weaker form of Lemma 2.62, with a worse constant, was given by Hardy and Wright [72, Theorem 188], while Lemmas 2.61 and 2.62 above were essentially proved by W. Schmidt [149, pp. 22–23]. One can also deduce these results by filling in the details in [10, p. 47].

Theorems similar to Lemmas 2.63 and 2.65 were given by Cusick and Mendès France [49]. Instead of considering $\sup_{q \geq 1} q\|q\alpha\|$, they studied

$$\limsup_{q \to \infty} q\|q\alpha\|,$$

which is somewhat more natural; see also Perron [124]. Chowla [45] proved in 1931 that $K(a\alpha/b) < 2ab(K(\alpha) + 1)^3$, a bound much weaker, however, than our Corollary 2.64.

The Maillet continued fraction construction continues to be of interest; see for example [1].

Finally, the real mapping $G \colon x \mapsto 1/x$ (mod 1) for $x \neq 0$ with $G(0) = 0$ is often called the *Gauss map*. It has fascinating dynamics, some of which may be studied in [48].[1] Figure 2.1 shows the map drawn on the torus and hints at subtlety. Of course, drawn in the real plane, it has singularities at $1/n$ for $n = 1, 2, \ldots$

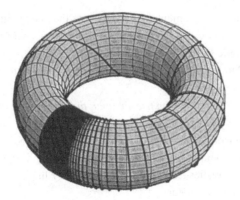

Figure 2.1 The Gauss map drawn on the torus.

---

[1] See also http://www.cecm.sfu.ca/organics/papers/corless/.

3

Metric theory of continued fractions

The examples we saw in Chapters 1 and 2 suggest that real numbers are arithmetically quite diverse. The theory of continued fractions as we have developed it allows us to recognise whether a given real number is rational or is a quadratic irrational; for the latter as well as for several transcendental numbers such as e, whose quotients follow a clear periodic pattern, we have precise knowledge of the quality of their rational approximations, as for instance in (2.40).

A standard counting argument, however, shows that the totality of such numbers is countable; hence they form a *subset of measure zero* of the reals. It is therefore reasonable to look into the arithmetic properties of other real numbers – in particular, of *almost all real numbers* (of course, in the sense of the usual Lebesgue measure $\mathcal{M}$).

The classical problems of *metric number theory* include determining the measure of the set of numbers that satisfy a given arithmetic property. In the context of continued fractions, for example, we may ask about the measure of the set of numbers whose 100th quotient a_{100} is exactly 100, or whose 100th convergent p_n/q_n satisfies $q_n < 10^{10}$. This is exactly the sort of question that we will address in this chapter.

Since shifting a real number α by an integer does not affect its arithmetic properties, we can always restrict our consideration to the case of real α between 0 and 1, so that we can write

$$\alpha = \cfrac{1}{a_1 + \cfrac{1}{a_2 + \cfrac{1}{a_3 + \cfrac{1}{a_4 + \ddots}}}} = [0; a_1, a_2, a_3, a_4, \ldots]. \qquad (3.1)$$

This will be our convention about α throughout this chapter and, for simplicity of notation, we may and will write $\alpha = [a_1, a_2, a_3, a_4, \dots]$.

3.1 Partial quotients of a number as functions of that number

The main task of this section is to express the partial quotients a_n in (3.1) as functions of α.

We call the interval $(0, 1)$, where our number α lives, the *interval of zeroth rank*. We are in effect studying the Gauss map mentioned at the end of the previous chapter. Because $a_1 = \lfloor 1/\alpha \rfloor$, we see that

$$a_1(\alpha) = k \qquad \text{for } \frac{1}{k+1} < \alpha \le \frac{1}{k}, \tag{3.2}$$

where $k \in \mathbb{N}$. In other words the only discontinuities of the function $a_1(\alpha)$ are at the points α where $1/\alpha$ is an integer, and $a_1(\alpha)$ is monotone increasing to ∞ as $\alpha \to 0$. The function $a_1(\alpha)$ is constant on the intervals $1/(k+1) < \alpha \le 1/k$, which we will call intervals of first rank.

Furthermore, the integral

$$\int_0^1 a_1(\alpha) \, d\alpha$$

diverges because of the divergence of the corresponding infinite series

$$\sum_{k=1}^{\infty} k \left(\frac{1}{k} - \frac{1}{k+1} \right) = \sum_{k=1}^{\infty} \frac{1}{k+1}.$$

Fixing $a_1(\alpha) = \lfloor 1/\alpha \rfloor = k$ and passing to the next quotient a_2, we see from a similar analysis that the function given by

$$a_2(\alpha) = l \qquad \text{for } \frac{1}{k+1/l} < \alpha \le \frac{1}{k+1/(l+1)}, \tag{3.3}$$

is constant on intervals which we will call the intervals of second rank. Any such interval is completely determined by the two conditions $a_1 = k$ and $a_2 = l$. Note that, as k increases, the corresponding intervals of second rank are arranged from left to right while the corresponding intervals of first rank are arranged from right to left.

Assuming that the intervals of rank n are defined and that the values

$$a_1(\alpha) = k_1, \quad a_2(\alpha) = k_2, \quad \dots, \quad a_n(\alpha) = k_n \tag{3.4}$$

are fixed, we write

$$\alpha = [a_1, a_2, \ldots, a_n, \alpha_{n+1}] = \frac{p_n \alpha_{n+1} + p_{n-1}}{q_n \alpha_{n+1} + q_{n-1}}, \tag{3.5}$$

where p_n, q_n, p_{n-1} and q_{n-1} are constant, as they are fully determined by $a_1, a_2, \ldots, a_n$. The complete quotient α_{n+1} ranges between 1 and ∞, so that α lies within the interval with endpoints

$$\frac{p_n + p_{n-1}}{q_n + q_{n-1}} \quad \text{and} \quad \frac{p_n}{q_n}.$$

As

$$\alpha - \frac{p_n}{q_n} = \frac{(-1)^n}{q_n(q_n \alpha_{n+1} + q_{n-1})},$$

α itself, viewed as a function of $\alpha_{n+1} > 1$, is monotone; hence $a_{n+1}(\alpha) = \lfloor \alpha_{n+1} \rfloor$ is monotone on the interval (of rank n)

$$I_n = \left(\frac{p_n}{q_n}, \frac{p_n + p_{n-1}}{q_n + q_{n-1}} \right). \tag{3.6}$$

Therefore, when α runs over the interval I_n, the corresponding $a_{n+1}(\alpha)$ assume successively the values $1, 2, 3, \ldots$, splitting I_n into a countable set of intervals of rank $n + 1$. The latter set is arranged from right to left if n is even and from left to right if n is odd.

To summarise, for each system of values (3.4) a unique interval (3.6) is assigned, and the function $a_{n+1}(\alpha)$ then takes all integer values from 1 to ∞ on the interval. In general, a system of values

$$a_{m_1}(\alpha) = k_1, \quad a_{m_2}(\alpha) = k_2, \quad \ldots, \quad a_{m_n}(\alpha) = k_n \tag{3.7}$$

determines a countable union of disjoint intervals.

The first question that we shall address in the framework of metric theory is determining the measure of the subset $\alpha \in (0, 1)$ for which $a_{n+1} = k$, a given positive integer.

Assuming that all partial quotients $a_1, \ldots, a_n$ are fixed as in (3.4), we know that our α belongs to the interval I_n in (3.6). The condition $a_{n+1} = \lfloor \alpha_{n+1} \rfloor = k$ implies that $k \le \alpha_{n+1} < k + 1$, so that the corresponding α in (3.5) belongs to the interval

$$I_{n+1}^{(k)} = \left[\frac{p_n k + p_{n-1}}{q_n k + q_{n-1}}, \frac{p_n(k+1) + p_{n-1}}{q_n(k+1) + q_{n-1}} \right).$$

Furthermore, the lengths of the intervals are

$$|I_n| = \left| \frac{p_n}{q_n} - \frac{p_n + p_{n-1}}{q_n + q_{n-1}} \right| = \frac{1}{q_n(q_n + q_{n-1})} = \frac{1}{q_n^2(1 + \lambda)}$$

and

$$|I_{n+1}^{(k)}| = \left| \frac{p_n k + p_{n-1}}{q_n k + q_{n-1}} - \frac{p_n(k+1) + p_{n-1}}{q_n(k+1) + q_{n-1}} \right|$$

$$= \frac{1}{(q_n k + q_{n-1})(q_n(k+1) + q_{n-1})}$$

$$= \frac{1}{q_n^2 k^2 (1 + \lambda/k)(1 + 1/k + \lambda/k)}$$

where $\lambda = q_{n-1}/q_n$. Therefore

$$\frac{|I_{n+1}^{(k)}|}{|I_n|} = \frac{1}{k^2} \frac{1 + \lambda}{(1 + \lambda/k)(1 + 1/k + \lambda/k)}$$

and, because

$$\frac{1 + \lambda}{1 + \lambda/k} \geq 1 \quad \text{and} \quad 1 + \frac{1}{k} + \frac{\lambda}{k} < 3,$$

we finally obtain

$$\frac{1}{3k^2} < \frac{|I_{n+1}^{(k)}|}{|I_n|} < \frac{2}{k^2}. \tag{3.8}$$

In other words, the interval of rank $n + 1$, where $a_{n+1} = k$, equals about $1/k^2$ of the underlying interval of rank n.

It is striking that this property, as well as the bounds in (3.8), are independent of the previous quotients $k_1, \ldots, k_n$ and are even independent of n. By summing the inequalities

$$\frac{1}{3k^2}|I_n| < |I_{n+1}^{(k)}| < \frac{2}{k^2}|I_n|$$

over all intervals I_n of rank n (that is, over all $k_1, \ldots, k_n$ ranging from 1 to ∞), and using the obvious equalities

$$\sum_n |I_n| = 1 \quad \text{and} \quad \sum_n |I_{n+1}^{(k)}| = \mathcal{M}(\{\alpha \in (0,1) : a_{n+1}(\alpha) = k\}),$$

we find that the contribution of the real numbers $\alpha \in (0,1)$, subject to the condition $a_{n+1}(\alpha) = k$, is between $1/(3k^2)$ and $2/k^2$ (and hence of magnitude $1/k^2$). This answers the question above – at least, to a first approximation.

3.2 Growth of partial quotients of a typical real number

Our knowledge is already sufficient to prove a metric result imposing infinitely many conditions on a given $\alpha \in \mathbb{R}$.

Theorem 3.1 *Let* $\psi \colon \mathbb{N} \to \mathbb{R}_{>0}$ *be an arbitrary positive function of a natural number n. Then, for almost all* $\alpha \in (0,1)$*, the inequality*

$$a_n = a_n(\alpha) \geq \psi(n) \tag{3.9}$$

holds infinitely often if the series

$$\sum_{n=1}^{\infty} \frac{1}{\psi(n)}$$

diverges. Conversely, for almost all $\alpha \in (0,1)$*, the inequality* (3.9) *holds at most finitely many times if the series converges.*

REMARK 3.2 In particular, taking $\psi(n) = M$, a positive constant, we deduce from Theorem 3.1 that the set

$$\mathcal{B}_M = \{\alpha \in (0,1) : a_n(\alpha) \leq M \text{ for all } n = 1, 2, \dots\}$$

has Lebesgue measure zero. Therefore, the set $\mathcal{B} = \bigcup_{M=1}^{\infty} \mathcal{B}_M$ of numbers all of whose partial quotients are bounded by some constant, which we introduced in Section 2.13, has Lebesgue measure zero as well (as does a countable union of such null sets).

Proof of Theorem 3.1 Let I_{m+n} denote an interval of rank $m+n$ whose points α satisfy

$$a_{m+j} = a_{m+j}(\alpha) < \psi(m+j) \qquad \text{for } j = 1, 2, \dots, n, \tag{3.10}$$

where we do not impose constraints on the quotients $a_1(\alpha), \dots, a_m(\alpha)$. As in Section 3.1, let $I_{m+n+1}^{(k)}$ denote the subinterval of I_{m+n} whose points satisfy $a_{m+n+1}(\alpha) = k$. Then, by the result in the previous section, we have

$$|I_{m+n+1}^{(k)}| = \mathcal{M}(I_{m+n+1}^{(k)}) > \frac{1}{3k^2}\mathcal{M}(I_{m+n}) = \frac{1}{3k^2}|I_{m+n}|.$$

Hence

$$\mathcal{M}\left(\bigcup_{k<M} I_{m+n+1}^{(k)}\right) = \sum_{k=1}^{\infty} |I_{m+n+1}^{(k)}| - \sum_{k\geq M} |I_{m+n+1}^{(k)}|$$

$$< |I_{m+n}| - \frac{1}{3}|I_{m+n}| \sum_{k\geq M} \frac{1}{k^2}$$

$$< |I_{m+n}|\left(1 - \frac{1}{3} \int_{M+1}^{\infty} \frac{dx}{x^2}\right)$$

$$= \left(1 - \frac{1}{3(M+1)}\right)|I_{m+n}|,$$

where we set $M = \psi(m+n+1)$. Letting $E_{m,n}$ denote the union of all intervals

I_{m+n} of rank $m + n$, subject to the constraints (3.10), and summing the latter estimate over all such intervals, we obtain

$$\mathcal{M}(E_{m,n+1}) < \left(1 - \frac{1}{3(\psi(m + n + 1) + 1)}\right)\mathcal{M}(E_{m,n}).$$

Iteration of this expression yields

$$\mathcal{M}(E_{m,n}) < \mathcal{M}(E_{m,1}) \prod_{j=2}^{n}\left(1 - \frac{1}{3(\psi(m + j) + 1)}\right).$$

For the divergent series $\sum_{n=1}^{\infty} 1/\psi(n)$, the series

$$\sum_{j=2}^{\infty} \frac{1}{3(\psi(m + j) + 1)}$$

diverges as well since if $\sum_n |a_n|$ diverges then so does $\sum_n |a_n|/(1+|a_n|)$. Thence, see [157, Theorem 7.28], the product

$$\prod_{j=2}^{n}\left(1 - \frac{1}{3(\psi(m + j) + 1)}\right)$$

tends to 0 as $n \to \infty$. This implies that, for any fixed m, we have

$$\mathcal{M}(E_{m,n}) \to 0 \qquad \text{as } n \to \infty,$$

and hence the set $E_m = \bigcap_{n=1}^{\infty} E_{m,n} = \{\alpha \in (0, 1) : a_n(\alpha) < \psi(n) \text{ for all } n > m\}$ has measure zero as well. Finally, if for a real number α the inequality (3.9) is satisfied at most finitely many times then $\alpha \in E_m$ for sufficiently large m. Hence α is also in the set $E = \bigcup_{m=1}^{\infty} E_m$, which has measure zero, as it is a countable union of measure-zero sets. This establishes the first part of our theorem.

Now assume that the series $\sum_{n=1}^{\infty} 1/\psi(n)$ converges; in particular, for all $n \geq n_0$ we have $\psi(n) > 2$, and the series

$$\sum_{n=n_0}^{\infty} \frac{1}{\psi(n) - 1}$$

converges as well. Take an arbitrary $n \geq n_0$. Let I_n be an interval of rank n and $I_{n+1}^{(k)} \subset I_n$ the embedded interval of rank $n + 1$ subject to the condition $a_{n+1} = k$. The upper bound in (3.8) implies that

$$\sum_{k \geq \psi(n+1)} |I_{n+1}^{(k)}| < 2|I_n| \sum_{k \geq \psi(n+1)} \frac{1}{k^2} \leq 2|I_n| \int_{\psi(n+1)-1}^{\infty} \frac{dx}{x^2}$$

$$= \frac{2|I_n|}{\psi(n + 1) - 1}.$$

Writing $E'_n = \{\alpha \in (0, 1) : a_n(\alpha) \geq \psi(n)\}$ and summing the resulting inequality over all intervals of rank n, we find that

$$\mathcal{M}(E'_{n+1}) < \frac{2}{\psi(n+1) - 1},$$

and so the series $\sum_{n=n_0}^{\infty} \mathcal{M}(E'_n)$ converges.

Thus, if E' is the set of $\alpha \in (0, 1)$ for which $a_n \geq \phi(n)$ holds infinitely often (so that those α belong to infinitely many sets E'_n) then $\mathcal{M}(E') = 0$. (The latter is a standard fact from metric set theory: the set E' is contained in $\bigcup_{n=m}^{\infty} E'_n$ for any $m \geq n_0$, and the measure of the latter can be made arbitrarily small because it is bounded by a tail of the convergent series $\sum_{n=n_0}^{\infty} \mathcal{M}(E'_n)$; see [157, Chapter 6].) This proves the second part of the theorem. $\quad\square$

3.3 Approximation of almost all real numbers by rationals

Theorem 1.38 tells us that for any irrational α the inequality

$$\left| \alpha - \frac{p}{q} \right| < \frac{1}{q^2}$$

has infinitely many solutions in $p \in \mathbb{Z}$ and $q \in \mathbb{N}$.

In the case where $\alpha \in \mathbb{R}$ with all partial quotients bounded (for example, for quadratic irrationalities α), the function $\phi(q) = 1/q^2$ gives the best approximation order of α by rationals (see the definition after (1.14)); this follows from the bounds in (2.17). In Section 2.12 we examined the best approximation order of e (and of related numbers with particular patterns in their continued fractions). In both these examples the numbers form sets of measure zero.

The aim of this section is to quantify this by proving the following theorem.

Theorem 3.3 *Let $\phi(x)$ be a positive continuous function of a positive real number x such that $x^2\phi(x)$ is non-decreasing. Then, for almost all $\alpha \in \mathbb{R}$, the inequality*

$$\left| \alpha - \frac{p}{q} \right| < \phi(q) \tag{3.11}$$

has infinitely many solutions in $p \in \mathbb{Z}$ and $q \in \mathbb{N}$, if (for some $c > 0$) the improper integral

$$\int_c^{\infty} x\phi(x)\,\mathrm{d}x \tag{3.12}$$

diverges. Conversely, if the integral converges then for almost all $\alpha \in \mathbb{R}$ the inequality has only finitely many solutions in integers p and $q > 0$.

We need two preparatory lemmas.

Lemma 3.4 *For a positive real M, consider the n-fold integral*

$$T_n(M) = \int \cdots \int_{\substack{x_1, x_2, \ldots, x_n \geq 1 \\ x_1 x_2 \cdots x_n \geq M}} \frac{dx_1 \, dx_2 \cdots dx_n}{x_1^2 x_2^2 \cdots x_n^2} = \int \cdots \int_{\substack{0 < y_1, y_2, \ldots, y_n \leq 1 \\ y_1 y_2 \cdots y_n \leq 1/M}} dy_1 \, dy_2 \cdots dy_n \qquad (3.13)$$

(the second integral is the result of the change of variables $x_j = 1/y_j$, for $j = 1, \ldots, n$, in the first integral). Then $T_n(M) = 1$ for $0 < M \leq 1$, and

$$T_n(M) = \frac{1}{M} \sum_{k=0}^{n-1} \frac{\log^k M}{k!} \qquad \text{for } M \geq 1. \qquad (3.14)$$

Proof If $M \leq 1$ then the second integral in (3.13) represents the volume of the *n*-dimensional unit cube, so that $T_n(M) = 1$.

Take $M \geq 1$. Then

$$T_1(M) = \int_0^{1/M} dy = \frac{1}{M};$$

in other words the formula (3.14) holds for $n = 1$. Assuming it holds for $n - 1$ and splitting the integration over y_n in (3.13) into the two intervals $(0, 1/M]$ and $(1/M, 1]$, we get

$$T_n(M) = \int_0^1 T_{n-1}(My_n) \, dy_n$$

$$= \int_0^{1/M} T_{n-1}(My_n) \, dy_n + \int_{1/M}^1 T_{n-1}(My_n) \, dy_n$$

$$= \int_0^{1/M} dy_n + \frac{1}{M} \int_1^M T_{n-1}(y) \, dy$$

$$= \frac{1}{M} + \frac{1}{M} \int_1^M \sum_{k=0}^{n-2} \frac{\log^k y}{k!} \frac{dy}{y} = \frac{1}{M} + \frac{1}{M} \sum_{k=0}^{n-2} \frac{\log^{k+1} y}{(k+1)!},$$

which proves (3.14) for *n*. □

Lemma 3.5 *There exists an absolute constant $B > 0$ such that, for almost all $\alpha = [a_0; a_1, a_2, \ldots] \in \mathbb{R}$ and sufficiently large n, the inequality*

$$a_1 a_2 \cdots a_n < e^{Bn} \qquad (3.15)$$

holds. (More explicitly, one can use (3.15) with $B = 3$.)

Proof For each $n = 1, 2, \ldots$ and a real number $M \geq 1$, introduce the set $E_n(M)$ of $\alpha \in (0, 1)$ for which the inequality $a_1 a_2 \cdots a_n \geq M$ holds. This set is clearly a finite union of certain intervals of rank n; we know from Section 3.1 that the length of any such interval is

$$\left| \frac{p_n}{q_n} - \frac{p_n + p_{n-1}}{q_n + q_{n-1}} \right| = \frac{1}{q_n(q_n + q_{n-1})} < \frac{1}{q_n^2} < \frac{1}{(a_1 a_2 \cdots a_n)^2},$$

where we have used the lower bound from (2.18). Thus,

$$\mathcal{M}(E_n(M)) < \sum_{\substack{a_1, a_2, \ldots, a_n \geq 1 \\ a_1 a_2 \cdots a_n \geq M}} \frac{1}{(a_1 a_2 \cdots a_n)^2}. \qquad (3.16)$$

To estimate the sum on the right-hand side of (3.16), write

$$\prod_{j=1}^{n} \frac{1}{a_j^2} = \prod_{j=1}^{n} \left(1 + \frac{1}{a_j} \right) \frac{1}{a_j(a_j + 1)}$$

$$\leq 2^n \prod_{j=1}^{n} \frac{1}{a_j(a_j + 1)} = 2^n \prod_{j=1}^{n} \int_{a_j}^{a_j+1} \frac{\mathrm{d}x_j}{x_j^2}.$$

Hence

$$\sum_{a_1 a_2 \cdots a_n \geq M} \prod_{j=1}^{n} \frac{1}{a_j^2} \leq 2^n T_n(M),$$

where $T_n(M)$ is given in (3.13). By Lemma 3.4 we can extend (3.16) to

$$\mathcal{M}(E_n(M)) < \frac{2^n}{M} \sum_{k=0}^{n-1} \frac{\log^k M}{k!}.$$

Now, choosing $M = e^{Bn}$, where $B > 1$ is constant, using

$$\frac{\log^k M}{k!} = \frac{(Bn)^k}{k!} < \frac{(Bn)^n}{n!} \qquad \text{for } k = 0, 1, \ldots, n,$$

and the estimate $n! > (n/e)^n$ obtained from exponentiating the lower bound in (2.37), we obtain

$$\mathcal{M}(E_n(e^{Bn})) < e^{(\log 2 - B)n} n \frac{(Bn)^n}{n!} < n e^{(1 + \log 2 + \log B - B)n}$$

$$< e^{(6/5 + \log 2 + \log B - B)n}.$$

In particular, the series $\sum_{n=1}^{\infty} \mathcal{M}(E_n(e^{3n}))$ converges. But this means that $\mathcal{M}\left(\cap_{m=1}^{\infty} \cup_{n \geq m} (E_n(e^{3n})) \right) = 0$ and, except for a set of measure zero, each real number belongs to at most finitely many sets $E_n(e^{3n})$. In other words, for almost all $\alpha \in (0, 1)$ and sufficiently large n, one has

$$a_1 a_2 \cdots a_n < e^{3n},$$

which is the required inequality. $\qquad \square$

REMARK 3.6   Later we will see that a much stronger result is true, namely, that

$$\lim_{n\to\infty} \sqrt[n]{a_1 a_2 \cdots a_n} = \prod_{m=1}^{\infty}\left(1 + \frac{1}{m(m+2)}\right)^{\log_2 m}$$

$$= 2.685\,452\,001\,065\,306\,445\,309\,714\,835\,481\ldots \quad (3.17)$$

for almost all $\alpha \in \mathbb{R}$.

Lemma 3.7   *There exists an absolute constant $C > 1$ such that, for almost all $\alpha \in \mathbb{R}$ and sufficiently large n, we have*

$$q_n(\alpha) < e^{Cn}. \quad (3.18)$$

REMARK 3.8   From the results of Section 1.3 we already know that $q_n(\alpha) \geq F_n$, where F_n is the nth Fibonacci number, holds for *any* real number α; thus Lemma 3.7 implies that

$$\frac{1 + \sqrt{5}}{2} \leq \sqrt[n]{q_n(\alpha)} \leq C,$$

for almost all real α and sufficiently large n. In fact, a more delicate analysis [81, 98] (far beyond the scope of our exposition here) shows that for almost all $\alpha \in \mathbb{R}$ the following limit exists:

$$\lim_{n\to\infty} \sqrt[n]{q_n(\alpha)} = \exp\left(\frac{\pi^2}{12 \log 2}\right) = 3.275\,822\,918\,721\,811\ldots \quad (3.19)$$

Proof of Lemma 3.7   This follows immediately from Lemma 3.5 and the upper bound in (2.18), with the implied constant

$$C = B + \log \frac{1 + \sqrt{5}}{2}. \qquad \square$$

EXERCISE 3.9 (Lochs' theorem [101])   Using (3.19), show the following: if $n = n(d)$ is the exact number of partial quotients of the continued fraction for $\alpha \in [0, 1)$ that can be obtained from the first d decimals of α then

$$\lim_{d\to\infty} \frac{n}{d} = \frac{6 \log 2 \log 10}{\pi^2} = 0.970\,270\,114\,392\,033\,925\,74\ldots$$

for almost all α.

For example, knowing the first 1000 decimal digits of π allows one to compute the first 968 partial quotients of its continued fraction.

Proof of Theorem 3.3   First consider the case when the integral (3.12) diverges. Define $f(y) = e^{2Cy}\phi(e^{Cy})$, which is a non-decreasing function by the hypothesis, where $C > 1$ is the constant from Lemma 3.7. Then the integral

$$\int_c^M f(y)\,dy = \frac{1}{C}\int_c^M e^{Cy}\phi(e^{Cy})\,d(e^{Cy}) = \frac{1}{C}\int_{\exp(cC)}^{\exp(cM)} x\phi(x)\,dx,$$

where $M > c$, increases unboundedly as $M \to \infty$. It follows that the series $\sum_{n\geq c} f(n)$ diverges, and Theorem 3.1 implies that, for almost all $\alpha \in (0,1)$, the inequality $a_{n+1}(\alpha) \geq 1/f(n)$ holds for infinitely many indices n. This inequality in turn yields

$$\left|\alpha - \frac{p_n}{q_n}\right| \leq \frac{1}{q_n q_{n+1}} \leq \frac{1}{a_{n+1}q_n^2} \leq \frac{f(n)}{q_n^2}. \tag{3.20}$$

At the same time, for almost all α and sufficiently large n we have $q_n < e^{Cn}$ by Lemma 3.7, which is equivalent to $n > (\log q_n)/C$. Substituting this into (3.20) we obtain

$$\left|\alpha - \frac{p_n}{q_n}\right| < \frac{f((\log q_n)/C)}{q_n^2} = \phi(q_n),$$

which holds for almost all real numbers α and infinitely many indices n. This gives the required infinitude of integer solutions of (3.11).

Now suppose that the integral (3.12) converges. Then the series $\sum_{n\geq c} n\phi(n)$ converges. Let E_q denote the set of α in $(0,1)$ that satisfy (3.11). The set E_q is clearly a disjoint union of intervals, each of length $2\phi(q)$, whose centres are located at $1/q, 2/q, \ldots, (q-1)/q$, and of the two additional intervals $(0, \phi(q))$ and $(1 - \phi(q), 1)$. Thus $M(E_q) \leq 2q\phi(q)$ (strict inequality only occurs when $q\phi(q) > 1/2$), so that the series $\sum_{q\geq c} M(E_q)$ converges. This implies, as we have already seen several times, that almost all $\alpha \in (0,1)$ can belong only to a finite number of sets E_n. In other words, for almost all α and for q sufficiently large the inequality

$$\left|\alpha - \frac{p}{q}\right| \geq \phi(q)$$

holds. This proves the second part of the theorem.   □

Before we turn to the expected size of the coefficients in a continued fraction, we should remark that Lebesgue measure is only one way of quantifying smallness. This is made clear by the striking exercise that follows.

EXERCISE 3.10   Show that the set of Liouville numbers, all transcendental, has measure zero but nonetheless is large in the sense of Baire category; indeed the complement of the Liouville numbers is a nowhere dense F_σ-set (a countable union of closed sets).

3.4 Gauss–Kuzmin statistics

Writing

$$\alpha_n = \alpha_n(\alpha) = [a_n; a_{n+1}, a_{n+2}, \dots], \qquad n = 0, 1, 2, \dots,$$

for the complete quotients of $\alpha = [0; a_1, a_2, \dots] \in (0, 1)$, define

$$x_n(\alpha) = \alpha_n - \lfloor \alpha_n \rfloor = \alpha_n - a_n, \qquad n = 0, 1, 2, \dots$$

Then, clearly, $x_0(\alpha) = \alpha$ and $0 \le x_n(\alpha) < 1$ for $n = 0, 1, 2, \dots$ We will let $\mu_n(x)$ denote the measure of the set $\{\alpha \in (0, 1) : x_n(\alpha) < x\}$, where $0 \le x \le 1$. Then $\mu_0(x) = x$ and simple analysis shows that

$$\mu_n(x) = \sum_{k=1}^{\infty} \left(\mu_{n-1}\left(\frac{1}{k}\right) - \mu_{n-1}\left(\frac{1}{k+x}\right) \right), \tag{3.21}$$

$$0 \le x \le 1, \quad n = 1, 2, \dots$$

Indeed, because of the obvious relation $x_{n-1} = 1/(a_n + x_n)$, the inequality $x_n < x$ holds iff

$$\frac{1}{k+x} < x_{n-1} \le \frac{1}{k}$$

is satisfied for an integer $k \ge 1$; passing to measures, we obtain (3.21).

The functional equation (3.21) was known to Gauss, and probably formed the grounds for his claim in a letter to Laplace of having found a proof of

$$\lim_{n \to \infty} \mu_n(x) = \log_2(1 + x), \qquad 0 \le x \le 1.$$

Indeed, the function $\mu(x) = c \log(1 + x)$ for any $c \in \mathbb{R}$ satisfies the equation (3.21) 'in the limit',

$$\mu(x) = \sum_{k=1}^{\infty} \left(\mu\left(\frac{1}{k}\right) - \mu\left(\frac{1}{k+x}\right) \right), \qquad 0 \le x \le 1;$$

the boundary condition $\mu(1) = 1$ then gives $c = 1/\log 2$. Besides this observation, Gauss indicated that it would be desirable to estimate the deviation $\mu_n(x) - \log_2(1 + x)$ for large n, a problem he could not solve himself.

Justification of the claim of Gauss, as well as an answer to his problem, was given by R. Kuzmin [91] only in 1928. One year later, but independently, another proof with a better estimate of the deviation was given by P. Lévy [97]. Indeed, a good proof needed the development of modern probability theory.

Theorem 3.11 (Gauss, Kuzmin, Lévy) *There exist absolute positive constants λ and C such that*

$$|\mu_n(x) - \log_2(1 + x)| < C e^{-\lambda n} \qquad \text{for } 0 \le x \le 1.$$

Although elementary, the proofs of the theorems involve many technicalities, so we refer the reader to the original sources [91, 97] as well as to the monograph [82] of Khintchine.

Theorem 3.11 can be applied to the problem with which we started this chapter, to determine the measure of the subset $\alpha \in (0, 1)$ for which $a_{n+1}(\alpha) = k$. Because the condition is equivalent to $1/(k + 1) < x_n(\alpha) \le 1/k$, we deduce that

$$\mathcal{M}(\{\alpha \in (0, 1) : a_{n+1}(\alpha) = k\}) = \mu_n\left(\frac{1}{k}\right) - \mu_n\left(\frac{1}{k + 1}\right),$$

so that

$$\left|\mathcal{M}\left(\{\alpha \in (0, 1) : a_{n+1}(\alpha) = k\} - \log_2\left(1 + \frac{1}{k(k + 2)}\right)\right)\right| < \frac{Ce^{-\lambda n}}{k(k + 1)}. \qquad (3.22)$$

In particular, our rough estimates in Section 3.1 can be now replaced with the sharp asymptotics

$$\lim_{n \to \infty} \mathcal{M}(\{\alpha \in (0, 1) : a_{n+1}(\alpha) = k\}) = \log_2\left(1 + \frac{1}{k(k + 2)}\right). \qquad (3.23)$$

For example, the measure of the set of reals, for which $a_{n+1}(\alpha) = 1$, tends to $2 - \log_2 3 = 0.415\,037\,499\,278\,843\,818\,546\,261\ldots$ as $n \to \infty$.

Of course, Theorem 3.11 was designed to solve much harder problems in metric number theory. In particular, the following theorem due to Khintchine [80, 82] holds.

Theorem 3.12 (Khintchine)   *Let* $f : \mathbb{N} \to (0, \infty)$ *be a function of a natural argument m. Suppose that there exist two positive constants C and δ such that* $f(m) < Cm^{1/2 - \delta}$ *for* $m = 1, 2, \ldots$ *Then for all* $\alpha \in (0, 1)$*, with exception of a set of measure zero, the following limit holds:*

$$\lim_{n \to \infty} \sum_{j=1}^{n} f(a_j) = \sum_{m=1}^{\infty} f(m) \log_2\left(1 + \frac{1}{m(m + 2)}\right). \qquad (3.24)$$

Taking $f(m) = \log m$ in Theorem 3.12 implies the geometric-mean asymptotics (3.17). The theorem, however, cannot be applied to the arithmetic mean $(a_1 + \cdots + a_n)/n$, as the corresponding choice $f(m) = m$ does not satisfy the hypothesis. However, a more elementary argument – namely, that in Theorem 3.1 – implies that, almost everywhere, for infinitely many n we have $a_n > n \log n$. Therefore

$$\frac{1}{n} \sum_{j=1}^{n} a_j > \frac{a_n}{n} > \log n,$$

and hence for almost all α the limit of $(a_1 + \cdots + a_n)/n$ as $n \to \infty$ does not exist.

The choice

$$f(m) = \begin{cases} 1 & \text{if } m = k, \\ 0 & \text{if } m \neq k \end{cases}$$

in Theorem 3.12, where k is a fixed positive integer, allows one to count $s_n(k) = \sum_{j=1}^{n} f(j)$, the number of appearances of k in $a_1, \ldots, a_n$. Then $s_n(k)/n$ characterises the density of k in the set of the first n partial quotients, while the limit

$$d(k) = \lim_{n \to \infty} \frac{s_n(k)}{n},$$

if it exists, represents the density of k in the set of all partial quotients. In this case, Theorem 3.12 implies the existence of the limit for almost all α, as well as providing its value:

$$d(k) = \log_2\left(1 + \frac{1}{k(k+2)}\right) \qquad \text{for } k = 1, 2, \ldots \tag{3.25}$$

Notes

Although the sets $\mathcal{B}_M$ in Remark 3.2 have measure zero, the set of denominators *(continuants)* $Q_M \subset \mathbb{N}$ of all convergents of the numbers in $\mathcal{B}_M$ is expected to coincide with $\mathbb{N}$ for all $M \geq 5$. This is known as Zaremba's conjecture [172], and we take it up again in Chapter 6. Quite recently, progress on the problem has been reported by Bourgain and Kontorovich [33]: for $M > 2189$, they showed that

$$\lim_{n \to \infty} \frac{\#\{q \in Q_M : 1 \leq q \leq n\}}{n} = 1.$$

The constants on the right-hand sides of (3.17) and (3.19) are known as Khintchine's and Lévy's constants, respectively [58].

An amusing self-referent question is whether Khintchine's constant (3.17) obeys the Gauss–Kuzmin statistics of (3.23) or (3.25). The computation of over 7000 digits of Khintchine's constant – and related quantities – as described in [9], suggests that it does. That paper also discusses explicit constructions of numbers for which the various Khintchine means hold.

Likewise, in Figure 3.1 we show a histogram of the first 100 million terms of the continued fraction of π, as computed by Bickford which are accessible

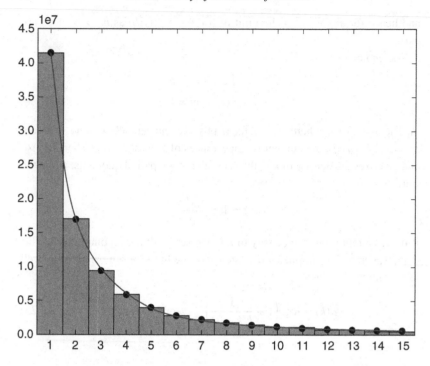

Figure 3.1 The expected and computed values of the Gauss–Kuzmin statistics for the first 100 million terms of the continued fraction for π.

at http://neilbickford.com/picf.htm. In Figure 3.2 we show a histogram of the error. The agreement is remarkable. Of course, nothing is proved!

Current information on the computation of the continued fraction for π is available at http://mathworld.wolfram.com/PiContinuedFraction.html. As of mid 2013, the largest terms found (Sloane's sequence A033089) out of the first 15 billion terms are

$$3, 7, 15, 292, 436, 20\,776, 78\,629, 179\,136, 528\,210, 12\,996\,958, 878\,783\,625,$$
$$5\,408\,240\,597, 5\,916\,686\,112, 9\,448\,623\,833.$$

The term $20\,776$ was found by Euler.

EXERCISE 3.13 Compute the first $10\,000$ partial quotients of your favourite real number (e.g., $\zeta(3)$, $\log 2$ etc.) and compare the distribution of the partial quotients with the statistical measures discussed in this chapter.

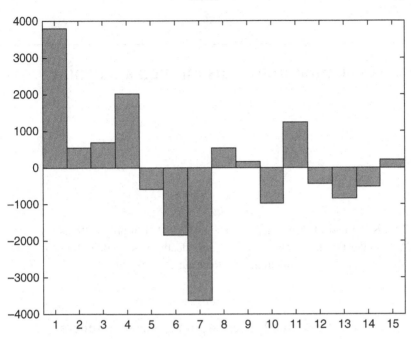

Figure 3.2 The error between the expected and computed values of the Gauss–Kuzmin statistics for the first 100 million terms of the continued fraction for π.

EXERCISE 3.14   Estimate the value of the so-called 'Khintchine harmonic mean' defined by

$$\lim_{n\to\infty} \frac{n}{a_1^{-1} + \cdots + a_n^{-1}}$$

for almost all real numbers. (A method for obtaining a high-precision answer for this and other such means is to be found in [9].)

4

Quadratic irrationals through a magnifier

This is the first of three chapters which originated in presentations given by Alf van der Poorten a few years before his death. As such they should be read like informal lectures and mined for their nuggets of gold.

4.1 Continued fractions of algebraic numbers

As we might expect by now, there is much more we can say about the continued fractions of quadratic irrationalities. First we look further in a more general way at algebraic numbers.

In spite of the expected unbounded behaviour of the continued fraction expansion of an *algebraic* non-quadratic irrational, there is a simple algorithm to compute its expansion. Indeed, it is quite straightforward [42, 95, 141] to find the beginning of the expansion of a real root of a polynomial equation.

Example 4.1 We illustrate this for the polynomial $f(X) = X^3 - X^2 - X - 1$. Then f has one real zero, say α, where $1 < \alpha < 2$. So $a_0 = 1$ and $\alpha_1 = 1/(\alpha - a_0)$ is a zero of the polynomial $f_1(X) = -X^3 f(X^{-1} + a_0) = 2X^3 - 2X - 1$. One sees that $\lfloor \alpha_1 \rfloor = 1$, so $a_1 = 1$ and $f_2(X) = -X^3 f_1(X^{-1} + a_1)$ is given by $X^3 - 4X^2 - 6X - 2$. A little more subtly, it happens that $\lfloor \alpha_2 \rfloor = 5$ and so $f_3(X) = 7X^3 - 29X^2 - 11X - 1$; the integer part of its real zero α_3 is $a_3 = 4$. This yields Table 4.1. The algorithm should now be perfectly clear and it barely seems worth continuing, particularly as a glance at the tabulation shows it will soon become very unwieldy.

The quadratic case is different, owing to the critical fact that the coefficients and degree of the $f_n(X)$ are bounded.

Exercise 4.2 Compute the continued fraction of $a + \sqrt{1 + a^2}$ for $a = 1, 2, 3, \ldots$

n	a_n	$f_n(X)$
0	1	$X^3 - X^2 - X - 1$
1	1	$2X^3 - 2X - 1$
2	5	$X^3 - 4X^2 - 6X - 2$
3	4	$7X^3 - 29X^2 - 11X - 1$
4	2	$61X^3 - 93X^2 - 55X - 7$
5	305	$X^3 - 305X^2 - 273X - 61$
6	1	$83\,326X^3 - 92\,752X^2 - 610X - 1$
7	8	$10\,037X^3 - 63\,864X^2 - 157\,226X - 83\,326$
8	2	$289\,486X^3 - 748\,054X^2 - 177\,024X - 10\,037$
9	1	$1\,040\,413X^3 - 304\,592X^2 - 988\,862X - 289\,486$
10	4	$542\,527X^3 - 1\,523\,193X^2 - 2\,816\,647X - 1\,040\,413$
11	6	$1\,956\,361X^3 - 11\,039\,105X^2 - 4\,987\,131X - 542\,527$
12	14	$5\,299\,117X^3 - 73\,830\,597X^2 - 24\,175\,393X - 1\,956\,361$
13	3	$270\,431\,827X^3 - 1\,024\,448\,687X^2 - 148\,732\,317X - 5\,299\,117$
14	1	$2\,369\,874\,922X^3 - 1\,006\,234\,890X^2 - 1\,409\,437\,756X - 270\,431\,827$
15	13	$316\,229\,551X^3 - 3\,687\,717\,230X^2 - 6\,103\,389\,876X - 2\,369\,874\,922$

Table 4.1

In real life, a fine idea [18] is that applying Vincent's theorem (see Theorem 4.9 below) makes it easy to produce many partial quotients at once and to avoid determining the intermediate polynomials $f_n(X)$; see Table 4.2. Quite exceptionally, $\alpha^{-17} = 56 - 103\alpha^{-1}$. That is the reason we chose the polynomial $f(X)$; indeed, $q_4^3 f(p_4/q_4) = -1$.

In fact, the comparative analysis of algorithms in [35] suggests that there is an even more efficient algorithm for computing continued fractions for the zeros of analytic functions (which is what algebraic numbers are, of course). We discuss this algorithm in Section 8.3 below.

4.2 Quadratic irrationals revisited

In this section, we will (try to) disregard the knowledge we already have from Sections 2.8 and 2.9, in order to get a different view of quadratic irrational numbers. In particular, we will no longer assume they are *real*.

We say that a complex number ω is an *algebraic integer* if both its *trace* $u = \text{Tr}(\omega) = \omega + \overline{\omega}$ and *norm* $w = \text{Norm}(\omega) = \omega\overline{\omega}$ are ordinary (rational) integers. If ω is an algebraic integer then the corresponding quadratic equation

$$X^2 - uX + w = 0$$

has integer coefficients, and vice versa. Because ω is irrational its discriminant

n	a_n	p_n	q_n	$p_n^3 - p_n^2 q_n - p_n q_n^2 - q_n^3$
		1	0	1
0	1	1	1	−2
1	1	2	1	1
2	5	11	6	−7
3	4	46	25	61
4	2	103	56	−1
5	305	31 461	17 105	83 326
6	1	31 564	17 161	−10 037
7	8	283 973	154 393	289 486
8	2	599 510	325 947	−1 040 413
9	1	883 483	480 340	542 527
10	4	4 133 442	2 247 307	−1 956 361
11	6	25 684 135	13 964 182	5 299 117
12	14	363 711 332	197 745 855	−270 431 827
13	3	1 116 818 131	607 201 747	2 369 874 922
14	1	1 408 529 463	804 947 602	−316 229 551
15	13			

Table 4.2

$D = D(\omega) = (\omega - \overline{\omega})^2 = u^2 - 4w$ is not a rational square, and $\omega \in \mathbb{Q}(\sqrt{d})$ where d is the square-free part of the rational integer D.

EXERCISE 4.3　Show that the algebraic integers form a ring in $\mathbb{Q}(\sqrt{d})$.

Solution　First, we will check that a number $\alpha = (u + v\sqrt{d})/2 \in \mathbb{Q}(\sqrt{d})$ is an algebraic integer iff u, v are integers satisfying

$$u \equiv v \pmod 2 \qquad \text{if } d \equiv 1 \pmod 4,$$
$$u \equiv v \equiv 0 \pmod 2 \qquad \text{if } d \equiv 2, 3 \pmod 4. \tag{4.1}$$

If the above conditions are satisfied then, with the help of

$$\text{Tr}(\alpha) = u \in \mathbb{Z} \qquad \text{and} \qquad \text{Norm}(\alpha) = \frac{u^2 - dv^2}{4},$$

it is not difficult to see that α is an algebraic integer. Conversely, if α is an algebraic integer then its trace has to be an integer. Therefore

$$-dv^2 = 4\,\text{Norm}(\alpha) - u^2$$

is an integer as well; since d is square-free, the number v has to be an integer. Thus, $u^2 - dv^2 \equiv 0 \pmod 4$, and the congruence implies that conditions (4.1) are satisfied.

Writing

$$\theta = \begin{cases} \dfrac{1 + \sqrt{d}}{2} & \text{if } d \equiv 1 \pmod{4}, \\ \sqrt{d} & \text{if } d \equiv 2, 3 \pmod{4}, \end{cases}$$

we can restate the criterion (4.1) above in the form $\alpha = x + y\theta$ with x and y integers. Therefore the set of algebraic integers in $\mathbb{Q}(\sqrt{d})$ is the $\mathbb{Z}$-module $\mathbb{Z}[\theta]$. A routine check shows that both θ and θ^2 are algebraic integers, and so $\mathbb{Z}[\theta]$ is a ring. □

If α is an arbitrary (not necessarily integer) element of the same field $\mathbb{Q}(\sqrt{d})$, then scaling the above complex number ω (if required), we can set $\alpha = (\omega + A)/B$ where the positive integer B divides the norm:

$$\text{Norm}(\omega + A) = (\omega + A)(\overline{\omega} + A).$$

This last condition is a critical convention: indeed, B dividing the norm is equivalent to the $\mathbb{Z}$-module $\langle B, \omega + A \rangle_{\mathbb{Z}}$ being an ideal of the integral domain $\mathbb{Z}[\omega]$. (Recall that an *ideal* in a commutative ring is a subring which is closed under multiplication by any member of the ring.) To see this, it suffices to notice that

$$\omega(\omega + A) = -(w + uA + A^2) + (u + A)(\omega + A)$$

is in $\langle B, \omega + A \rangle_{\mathbb{Z}}$ iff B divides the norm $w + uA + A^2$.

Example 4.4 Writing

$$\alpha = \frac{\sqrt{-163} + 17}{21}$$

is less than ideal: it is *not admissible*. In fact,

$$\alpha = \frac{\sqrt{-7987} + 119}{147}.$$

If the quadratic integer ω is real, so that its discriminant $D = u^2 - 4w$ is positive, then we distinguish ω from its conjugate $\overline{\omega}$ by insisting that $\omega > \overline{\omega}$. One now says (see Section 2.8) that $\alpha = (\omega + A)/B$ is *reduced* if and only if $\alpha > 1$ but $-1 < \alpha < 0$.

If ω is imaginary then its discriminant $D = u^2 - 4w$ is negative. In this case, one says that α is *reduced* if and only if both $|\alpha + \overline{\alpha}| \leq 1$ and $\alpha\overline{\alpha} \geq 1$.

EXERCISE 4.5 Check that if a real number α is reduced then necessarily both $2A + u$ and B are positive and less than $\omega - \overline{\omega}$.

Now, assume that $\alpha = (\omega + A)/B$ is reduced. Write a_n for the integer part $\lfloor \alpha_n \rfloor$ of $\alpha_n = (\omega + A_n)/B_n$. Thus a_n is a partial quotient in the continued fraction expansion of α_n, and the first step in that expansion is

$$\alpha_n = \frac{\omega + A_n}{B_n} = a_n - \frac{\overline{\omega} + A_{n+1}}{B_n} = a_n - \overline{\rho}_{-n}; \qquad (4.2)$$

here $A_{n+1} = a_n B_n - A_n - u$. Then $-1 < \overline{\rho}_{-n} < 0$, because $-\overline{\rho}_{-n}$ is the fractional part of α_n. Now consider the *conjugate step*

$$\rho_{-n} = \frac{\omega + A_{n+1}}{B_n} = a_n - \frac{\overline{\omega} + A_n}{B_n} = a_n - \overline{\alpha}_n.$$

One sees that a_n, which began life as the integer part of α_n, is also the integer part of ρ_{-n}, and furthermore ρ_{-n} is reduced. It now follows that $\alpha_{n+1} = -1/\overline{\rho}_{-n} = (\omega + A_{n+1})/B_{n+1}$, the next complete quotient in the expansion, is also reduced.

Thus, the continued fraction expansion of a reduced quadratic irrational $\alpha_0 = (\omega + A_0)/B_0$ is a sequence (4.2) of steps $n = 0, 1, 2, \ldots$, where

$$A_n + A_{n+1} + u = a_n B_n, \qquad -B_n B_{n+1} = (\omega + A_{n+1})(\overline{\omega} + A_{n+1}),$$

$$\text{and} \qquad \alpha_{n+1} = \frac{\omega + A_{n+1}}{B_{n+1}}.$$

Here all the complete quotients α_n and all the 'remainders' ρ_{-n} are reduced quadratic irrationals.

Because the α_n are reduced, it follows that $\omega - \overline{\omega}$ bounds both $2A_n + u$ and B_n (cf. Exercise 4.5). Hence there are *only finitely many possibilities* for a step in the expansion, and so the expansion must be ultimately periodic.

EXERCISE 4.6 For discussion: 'finitely many' only means 'fewer than infinity', but here we have much more explicit information. Think of how one might obtain a good upper bound on the length of an ideal cycle in the domain $\mathbb{Z}[\omega]$, say as a function of $D = u^2 - 4w$ as $D \to \infty$.

Example 4.7 Table 4.3 displays all the steps in the expansion of $\alpha = \omega = \sqrt{46}$. Here we see $\omega = \sqrt{46}$ displaying its period of length $r = 12$. The convergents p_n/q_n that are also computed here provide the interesting identity $p_n^2 - 46 q_n^2 = (-1)^{n+1} B_{n+1}$. In particular, $24335^2 - 46 \times 3588^2 = 1$.

Now suppose that step $r - 1$ is the first step in the table to coincide with an earlier step. Then the period length of the expansion of α is at most r and, unless step $r - 1$ happens to coincide with step 0, the expansion will have a pre-period.

	n	p_n	q_n
$(\sqrt{46}+0)/1 \;=\; 6-(-\sqrt{46}+6)/1$	0	6	1
$(\sqrt{46}+6)/10 = 1-(-\sqrt{46}+4)/10$	1	7	1
$(\sqrt{46}+4)/3 \;=\; 3-(-\sqrt{46}+5)/3$	2	27	4
$(\sqrt{46}+5)/7 \;=\; 1-(-\sqrt{46}+2)/7$	3	34	5
$(\sqrt{46}+2)/6 \;=\; 1-(-\sqrt{46}+4)/6$	4	61	9
$(\sqrt{46}+4)/5 \;=\; 2-(-\sqrt{46}+6)/5$	5	156	23
$(\sqrt{46}+6)/2 \;=\; 6-(-\sqrt{46}+6)/2$	6	997	147
$(\sqrt{46}+6)/5 \;=\; 2-(-\sqrt{46}+4)/5$	7	2 150	317
$(\sqrt{46}+4)/6 \;=\; 1-(-\sqrt{46}+2)/6$	8	3 147	464
$(\sqrt{46}+2)/7 \;=\; 1-(-\sqrt{46}+5)/7$	9	5 297	781
$(\sqrt{46}+5)/3 \;=\; 3-(-\sqrt{46}+4)/3$	10	19 038	2 807
$(\sqrt{46}+4)/10 = 1-(-\sqrt{46}+6)/10$	11	24 335	3 588
$(\sqrt{46}+6)/1 \;= 12-(-\sqrt{46}+6)/1$	12		

Table 4.3

However, consider the continued fraction expansion of ρ_{-r+1}, recalling that it commences with the step

$$\frac{\omega + A_r}{B_{r-1}} = a_{r-1} - \frac{\overline{\omega} + A_{r-1}}{B_{r-1}} = a_{r-1} - \overline{\alpha}_{r-1}.$$

Because this expansion is the conjugate of the continued fraction expansion of α, it too must have a period of length at most r. Because it commences with the conjugate of the first repeated line and runs in the direction opposite to that of the expansion of α, it must be purely periodic: any putative pre-period of α would provide a post-period for ρ_{-r+1}, which would be absurd. Thus, the expansion of α is also purely periodic, so we have proved the part of the Euler–Lagrange theorem (Theorem 2.48) addressing reduced quadratic irrationalities.

Let C denote the integer part of ω. For $\alpha_0 = \omega + C - u$, step 0 is

$$\alpha_0 = \omega + C - u = 2C - u - (\overline{\omega} + C - u) = 2C - u - \overline{\rho}_0$$

and is *symmetric*, that is, unchanged under conjugation. Though it is more natural to expand ω rather than $\omega + C - u$, we choose the latter because, unlike ω, it certainly is reduced.

EXERCISE 4.8

(a) Observe in the case $\omega + C - u$ that the period must have a second symmetry (at any rate, if $r > 1$). Moreover, if $r = 2k$ is even then this symmetry is given by $\alpha_k = \rho_{-k}$ and if $r = 2k + 1$ is odd then $\alpha_k = \rho_{-k+1}$.

(b) (*Mathematics is the study of degeneracy.*) The degenerate case here is $r = 1$. Does claim (a) remain true in essence (as it certainly should) for $r = 1$?

(c) It is not true that every α has a symmetric period. Comment on the claim that the period of α has symmetries iff there is an n such that either

    (i) α_n has integral trace

or

    (ii) $\alpha_n \overline{\alpha}_n = -1$.

(d) Give examples illustrating the various claims just made.

Again, set $\alpha = [a_0; a_1, a_2, \dots]$ with convergents denoted $[a_0; a_1, \dots, a_n] = p_n/q_n$. Suppose that we are great supporters of the number α, so much so that no matter what number, β say, we are expanding we always compute

$$\beta = [a_0; a_1, \dots, a_n, \beta_{n+1}]$$

using the *wrong* partial quotients. We have

$$\beta_{n+1} = -\frac{q_{n-1}\beta - p_{n-1}}{q_n\beta - p_n},$$

so we can readily compute the α-complete quotients. What more can one say about them?

Theorem 4.9 (Vincent's theorem [162]) *Either the β_n all lie in the left-hand half of the unit circle once n is sufficiently large, or $\beta = \alpha$ and they all are greater than 1.*

So what?

Suppose α is a real quadratic irrational and consider the α_n, recalling that complete quotients are all greater than 1; their conjugates $\overline{\alpha}_n$ are the result of α having suffered the ignominy of being α-expanded. Hence, once n is large enough they all satisfy $-1 < \overline{\alpha}_n < 0$. In other words, the continued fraction process eventually *reduces* any real quadratic irrational. This completes the proof of the Euler–Lagrange theorem.

4.3 Units and Pell's equation

A *unit* in a real quadratic domain $\mathbb{Z}[\omega]$ is an algebraic integer whose reciprocal is an algebraic integer as well. If α is a unit then the relation $\alpha\alpha^{-1} = 1$ implies that $\text{Norm}(\alpha) = \pm 1$. Conversely, if $\text{Norm}(\alpha) = \alpha\overline{\alpha} = \pm 1$ then α is a unit, since

$\alpha^{-1} = \pm\overline{\alpha}$. Thus, the units in $\mathbb{Z}[\omega]$ form a multiplicative group of nonzero algebraic integers whose norm is either 1 or -1.

We next apply the Dirichlet box principle (see Section 1.4) and Dirichlet's theorem (Theorem 1.36) itself to show that real quadratic domains $\mathbb{Z}[\omega]$ contain *nontrivial* units, that is, elements different from ± 1 yet dividing 1. The periodicity of the continued fraction expansion of a real quadratic irrational is a corollary. This argument is independent of our earlier arguments.

Given ω with trace $u \in \mathbb{Z}$ and norm $w \in \mathbb{Z}$, it follows from Theorem 1.38 that there are infinitely many integers q such that $|q\omega - p| < 1/q$; whence, after multiplying by the conjugate and because $|\omega - p/q| < 1$, we have

$$|(q\omega - p)(q\overline{\omega} - p)| < \omega - \overline{\omega} + 1.$$

Again by the Dirichlet box principle, it follows that there is some integer k, with $|k| < \omega - \overline{\omega} + 1$, for which there are infinitely many pairs of integers (p, q) such that

$$|(q\omega - p)(q\overline{\omega} - p)| = k.$$

Yet again, it follows by the box principle that there is a pair among those pairs such that $p \equiv p' \pmod{k}$ and $q \equiv q' \pmod{k}$. Then

$$\frac{(q\omega - p)(q\overline{\omega} - p)}{(q'\omega - p')(q'\overline{\omega} - p')} = (x - \omega y)(x - \overline{\omega}y) = \pm 1$$

displays a unit $\alpha = x - \omega y$; here x and y are rational integers given by

$$x = \frac{pp' - upq' + wqq'}{k} \quad \text{and} \quad y = \frac{pq' - p'q}{k}.$$

EXERCISE 4.10   Verify, or correct, all these remarks.

It is often convenient to set

$$L = \begin{pmatrix} 1 & 0 \\ 1 & 1 \end{pmatrix}, \quad J = \begin{pmatrix} 0 & 1 \\ 1 & 0 \end{pmatrix}, \quad R = \begin{pmatrix} 1 & 1 \\ 0 & 1 \end{pmatrix},$$

whence

$$\begin{pmatrix} a & 1 \\ 1 & 0 \end{pmatrix} = R^a J = J L^a.$$

Thus a continued fraction expansion $[a_0; a_1, a_2, \dots]$ corresponds to an *RL-sequence* $R^{a_0} L^{a_1} R^{a_2} L^{a_3} R^{a_4} \dots$ It follows, for example, that a zero partial quotient is readily dealt with by the rule

$$[\dots, a, 0, b, \dots] = [\dots, a + b, \dots]. \tag{4.3}$$

Now let

$$A = \begin{pmatrix} 2 & 0 \\ 0 & 1 \end{pmatrix} \quad \text{and} \quad A' = \begin{pmatrix} 1 & 0 \\ 0 & 2 \end{pmatrix}.$$

Multiplying a continued fraction by 2 is the same as multiplying its RL-sequence on the left by A. However, to turn that product back into an RL-sequence we now need rules for commuting the matrix A through the sequence.

EXERCISE 4.11

(a) Verify that $AR = R^2A$, $ALR = RLA'$ and $AL^2 = LA$ and obtain the corresponding transition rules for A'.

(b) Define ω by $\omega^2 - \omega - 15 = 0$. Compute its continued fraction expansion, and thence that of $\sqrt{61}$.

Given that $x - \omega y$ is a unit, the matrix

$$N = \begin{pmatrix} x & -wy \\ y & x - uy \end{pmatrix} \tag{4.4}$$

has determinant ± 1 and hence decomposes as a product:

$$N = \begin{pmatrix} b_0 & 1 \\ 1 & 0 \end{pmatrix}\begin{pmatrix} b_1 & 1 \\ 1 & 0 \end{pmatrix} \cdots \begin{pmatrix} b_r & 1 \\ 1 & 0 \end{pmatrix}\begin{pmatrix} 0 & 1 \\ 1 & 0 \end{pmatrix};$$

there is a concluding zero in the top left element because $-wy > x$.

Theorem 4.12   *The continued fraction expansion of ω is given by*

$$\omega = [\overline{b_0, b_1, \ldots, b_r, 0}] = [b_0; \overline{b_1, \ldots, b_r + b_0}].$$

Proof   Indeed, suppose that $[\overline{b_0, b_1, \ldots, b_r, 0}] = \beta$; in other words, that $\beta = [b_0; b_1, \ldots, b_r, 0, \beta]$. Then, by matrix correspondence,

$$\beta \leftrightarrow N\begin{pmatrix} \beta & 1 \\ 1 & 0 \end{pmatrix} = \begin{pmatrix} \beta x - wy & x \\ \beta y + x - uy & y \end{pmatrix} \leftrightarrow \frac{\beta x - wy}{\beta y + x - uy}.$$

Thus $(\beta^2 - u\beta + w)y = 0$. Because the given unit is nontrivial we have $y \neq 0$ and so $\beta^2 - u\beta + w = 0$, which we said we would prove.   □

We can also see that the period has a symmetry. It is plain that NJL^u is a symmetric matrix, so it follows that the list

$$b_0, b_1, \ldots, b_{r-1}, b_r + u$$

must be symmetric. In particular b_0 is $\lfloor \omega \rfloor = C$, so $b_r = C - u$. Note that all this would not make sense if u were not a rational integer.

The reader should now be able to see why starting with $\omega + C - u$ painlessly yields a purely periodic expansion.

EXERCISE 4.13 Set $\alpha = (\omega + A)/B$.

(a) Given that $x - \omega y$ is a unit, find *integers* a and b such that $a - b\alpha$ is a unit.

(b) Next, construct the matrix

$$N = \begin{pmatrix} a & -w_\alpha b \\ b & a - u_\alpha b \end{pmatrix},$$

with $w_\alpha = \alpha\overline{\alpha}$ and $u_\alpha = \alpha + \overline{\alpha}$, and decompose it as a product of matrices

$$\begin{pmatrix} c_i & 1 \\ 0 & 1 \end{pmatrix}.$$

(c) Show that such decompositions do indeed yield a period for α, in complete analogy with the special case ω.

(d) For discussion: what we have done cannot be quite right. Distinguish the cases where α is reduced and where it is not reduced in your discussion. What remarks are needed to correct the argument?

Recall the recursion formula $(\omega + A_{n+1})(\overline{\omega} + A_{n+1}) = -B_n B_{n+1}$ and the distance formula (2.30) (in which we replace n with $n + 1$):

$$\alpha_1 \alpha_2 \cdots \alpha_{n+1} = \frac{(-1)^{n+1}}{p_n - q_n \alpha}.$$

Because $\alpha_n \overline{\alpha}_n = -B_{n-1}/B_n$ and $B_0 = B$, taking norms yields

$$B p_n^2 - (2A + u)p_n q_n + \frac{w + uA + A^2}{B} q_n^2 = (-1)^{n+1} B_{n+1}.$$

In particular, if $\alpha = \omega$ then $A = 0$ and $B = 1$, so that

$$(p_n - \omega q_n)(p_n - \overline{\omega} q_n) = p_n^2 - u p_n q_n + w q_n^2 = (-1)^{n+1} B_{n+1}.$$

But $\omega + C - u$, and therefore also ω, is periodic with period r iff $B_r = 1$, in which case $p_{r-1}^2 - u p_{r-1} q_{r-1} + w q_{r-1}^2 = (-1)^r$ and $x - \omega y = p_{r-1} - \omega q_{r-1}$ is a unit.

Thus, we have

Theorem 4.14 *The existence of a unit in $\mathbb{Z}[\omega]$ and the periodicity of the continued fraction expansion of the elements of $\mathbb{Z}[\omega]$ are equivalent.*

The equation

$$(x - \omega y)(x - \overline{\omega} y) = 1$$

is known as *Pell's equation*.

A_n	B_n	n	a_n	p_n	q_n	$p_n^2 - 62q_n^2$
				1	0	1
0	1	0	7	7	1	−13
7	13	1	1	8	1	2
6	2	2	6	55	7	−13
6	13	3	1	63	8	1
7	1	4	14	937	119	−13
7	13	5	1	1 000	127	2
6	2	6	6	6 937	881	−13
6	13	7	1	7 937	1 008	1
7	1	8	14	118 055	14 993	−13
7	13	9	1	125 992	16 001	2
6	2	10	6	874 007	110 999	−13
6	13	11	1	999 999	127 000	1
7	1	12	14	14 873 993	1 888 999	−13
7	13	13	1	15 873 992	2 015 999	2

Table 4.4

Example 4.15 In Table 4.4, $\omega = \sqrt{62}$ and we display only the necessary data. We see that $\omega = [7; \overline{1, 6, 1, 14}]$ and observe the *fundamental unit* $\eta = 63 - 8\omega$ and its powers $\eta^2 = 7\,937 - 1\,008\omega$ and $\eta^3 = 999\,999 - 127\,000\omega$.

Exercise 4.16 Notice that $\alpha = 8 - \omega$ has norm 2 and that plainly $\alpha^2 = 2\eta$. But $\beta = 7 - \omega$ has norm −13, yet $\beta^{2k}/13^k$ is never a unit for $k = 1, 2, \ldots$ Show this.

Consider integer matrices N of the form (4.4). Suppose that x and y are relatively prime, that is, $\gcd(x, y) = 1$, and that $\det N = \pm B$ with $B > 0$. Then N has a decomposition given by

$$N = \begin{pmatrix} x & -wy \\ y & x - uy \end{pmatrix} = \begin{pmatrix} x & x' \\ y & y' \end{pmatrix}\begin{pmatrix} 1 & A \\ 0 & B \end{pmatrix},$$

with integers x', y' such that $xy' - x'y = \pm 1$ and some integer A from the interval $0 \leq A < B$. In brief, the decomposition provides a correspondence between N and an ideal $\langle B, \omega + A \rangle_{\mathbb{Z}}$ of $\mathbb{Z}[\omega]$ and (this is the point) *this correspondence preserves multiplication of the matrices and of the ideals.*

Remark 4.17 We identify 2×2 matrices $k\gamma$ and γ for nonzero constants k (see the notes for Chapter 2); therefore, when multiplying matrices (or ideals) the relevant product is the one *after removal* of any factor common to all elements.

Exercise 4.18

(a) Show that if B is square-free then it divides the matrix N^2 iff B divides the discriminant $D = u^2 - 4w$.

(b) Show that if $B = 4$ then 8 divides the matrix N^3.

What's going on here? The secret of the ideal matrices lies in this: if B is small relative to x and y then one of the two factors of $x^2 - uxy + wy^2$ is small; let us suppose that $|x - \omega y|$ is small. But then the beginning of the continued fraction expansion of x/y must coincide with the initial terms of the expansion of ω. Suppose n is maximal, so that the convergent p_n/q_n of ω also is a convergent of x/y. Then one may think of the ideal $\langle B_{n+1}, \omega + A_{n+1} \rangle$ as the reduced ideal *nearest* to the unreduced ideal $\langle B, \omega + A \rangle$. In fact, if $|x - \omega y|$ is small enough ($2B < \omega - \overline{\omega}$ will certainly do) then necessarily $x/y = p_n/q_n$ is a convergent of ω by Theorem 2.37. In that case the decomposition of $N = N_n$ is precisely the expression obtained from the matrix correspondence:

$$\omega = [a_0; a_1, \ldots, a_n, (\omega + A_{n+1})/B_{n+1}]$$

$$\leftrightarrow \begin{pmatrix} p_n & -wq_n \\ q_n & p_n - uq_n \end{pmatrix} = \begin{pmatrix} a_0 & 1 \\ 1 & 0 \end{pmatrix} \begin{pmatrix} a_1 & 1 \\ 1 & 0 \end{pmatrix} \cdots \begin{pmatrix} a_n & 1 \\ 1 & 0 \end{pmatrix} \begin{pmatrix} 1 & A_{n+1} \\ 0 & B_{n+1} \end{pmatrix}.$$

EXERCISE 4.19   Show that the product of any two ideal matrices is again a matrix of that special form.

EXERCISE 4.20   Suppose $\eta = a + \omega b$ is a unit of $\mathbb{Z}[\omega]$ and set $\eta^n = a_n + \omega b_n$. If both $D(\omega) = u^2 - 4w$ and $b = b_1$ are odd, show that b_n is even iff 3 divides n.

4.4 Negative expansions and the many surprises of the number 163

We get a sequence of positive partial quotients, say $\{a_n\}_n$, of a simple continued fraction expansion by *underestimating* each successive complete quotient by its floor. We obtain

$$[a_0; a_1, a_2, a_3, a_4, a_5, \ldots] = a_0 + \cfrac{1}{a_1 + \cfrac{1}{a_2 + \cfrac{1}{a_3 + \cfrac{1}{a_4 + \cfrac{1}{a_5 + \ddots}}}}}.$$

If, instead, we define the partial quotients by *overestimating* the successive complete quotients by their ceiling, we obtain a *negative continued fraction* with partial quotients $\{b_n\}$, say. But a negative continued fraction is just a regular continued fraction with partial quotients of alternating sign (cf. (5.4) below):

$$[b_0; b_1, b_2, b_3, b_4, b_5, \ldots]^- = b_0 - \cfrac{1}{b_1 - \cfrac{1}{b_2 - \cfrac{1}{b_3 - \cfrac{1}{b_4 - \cfrac{1}{b_5 - \cdots}}}}}$$

$$= b_0 + \cfrac{1}{\overline{b}_1 + \cfrac{1}{b_2 + \cfrac{1}{\overline{b}_3 + \cfrac{1}{b_4 + \cfrac{1}{\overline{b}_5 + \cdots}}}}}$$

$$= [b_0; \overline{b}_1, b_2, \overline{b}_3, b_4, \overline{b}_5, \ldots].$$

Here, $\overline{b}$ is a convenient shorthand for $-b$.

Of course, formulae for evaluating continued fractions cannot know or care about the signs of the partial quotients. If one is so moved, one can change the sign of all succeeding partial quotients in an expansion by inserting the string $0, \overline{1}, 1, \overline{1}, 0$ into an expansion. Then, for example,

$$-\pi = [\overline{3}; \overline{7}, \overline{15}, \overline{1}, \overline{292}, \overline{1}, \overline{\ldots}]$$

$$= [\overline{3}; 0, \overline{1}, 1, \overline{1}, 0, 7, 15, 1, 292, 1, \ldots]$$

$$= [\overline{4}; 1, 6, 15, 1, 292, 1, \ldots]$$

(cf. Exercise 2.39).

Example 4.21 (Negation lemma)   The computation

$$-\beta = 0 + \overline{\beta},$$
$$-1/\beta = \overline{1} + (\beta - 1)/\beta,$$
$$\beta/(\beta - 1) = 1 + 1/(\beta - 1),$$
$$\beta - 1 = \overline{1} + \beta,$$
$$1/\beta = 0 + 1/\beta$$

shows that $-\beta = [0, \overline{1}, 1, \overline{1}, 0, \beta]$.

EXERCISE 4.22

(a) Explain why inserting the string $0, 1, \overline{1}, 1, 0$ into an expansion has the same effect as inserting the string $0, \overline{1}, 1, \overline{1}, 0$.
(b) Does the negation lemma above fully justify our insertion claim?
(c) Confirm the 'zeros are eaten' rule (4.3).

All this is enough to provide a succinct summary of just how a simple continued fraction expansion $[a_0; a_1, a_2, a_3, \ldots]$, with all the a_n positive, may be transformed into a negative continued fraction

$$[b_0; \overline{b}_1, b_2, \overline{b}_3, b_4, \ldots],$$

where the entries have alternating sign. In brief, one arranges the alternation of sign by alternately inserting the appropriate string, $0, \overline{1}, 1, \overline{1}, 0$ or $0, 1, \overline{1}, 1, 0$, between the first pair of consecutive partial quotients that have the same sign. One finds that $[a_0; a_1, a_2, a_3, a_4, \ldots]$ becomes the negative continued fraction

$$[a_0 + 1; \{2\}^{a_1-1}, a_2 + 2, \{2\}^{a_3-1}, a_4 + 2, \ldots]^-,$$

where $\{2\}^n$ denotes the n-times repetition of 2 (see Section 2.14).

Theorem 4.23 *Set $\omega = \sqrt{p}$, where $p \equiv 3$ (mod 4) is a prime number other than 3 with the property that $\mathbb{Q}(\omega)$ has class number $h(p) = 1$ (that is, the reduced elements of $\mathbb{Q}(\omega)$ make up just one cycle). Then $\frac{1}{3}(b_0 + b_1 + \cdots + b_{r-1}) - r$ is the number $h(-p)$ of distinct equivalence classes of quadratic forms having discriminant $-p$; here $[b_0, b_1, \ldots, b_{r-1}]$ is the (minimal) period of the negative continued fraction expansion of $\sqrt{p} + \lceil p \rceil$.*

Even if one does not understand what the theorem alleges, the incidental implication that the sum $b_0 + b_1 + \cdots + b_{r-1}$ must be divisible by 3 should astonish. Note that experimentally and conjecturally a majority of the primes $p = 4n + 3$ have class number 1.

COMMENT Those bizarre strings of 2s led van der Poorten to start off with quite negative feelings about negative continued fractions. But eventually he learned not to underestimate the usefulness of overestimation; he first heard Theorem 4.23 in the course of Hirzebruch's Mordell Lecture at Cambridge, UK in 1975; it can be found in [168, Satz 3, p. 136].

Example 4.24 A reasonably hefty example may be helpful. Take $p = 163$ and set $\omega = \sqrt{163}$; note that $\lfloor \omega \rfloor = 12$. Then

$$(\omega + 12)/1 = 24 - (\omega + 12)/1,$$
$$(\omega + 12)/19 = 1 - (\omega + 7)/19,$$

$$(\omega + 7)/6 = 3 - (\omega + 11)/6,$$
$$(\omega + 11)/7 = 3 - (\omega + 10)/7,$$
$$(\omega + 10)/9 = 2 - (\omega + 8)/9,$$
$$(\omega + 8)/11 = 1 - (\omega + 3)/11,$$
$$(\omega + 3)/14 = 1 - (\omega + 11)/14,$$
$$(\omega + 11)/3 = 7 - (\omega + 10)/3,$$
$$(\omega + 10)/21 = 1 - (\omega + 11)/21,$$
$$(\omega + 11)/2 = 11 - (\omega + 11)/2.$$

Thus

$$\omega + 12 = \overline{[24, 1, 3, 3, 2, 1, 1, 7, 1, 11, 1, 7, 1, 1, 2, 3, 3, 1]}.$$

Exercise 4.25

(a) List the reduced elements $(\omega + A)/B$, $\omega^2 - 163 = 0$ and confirm that each reduced element appears in the computation above and thus that $h(163) = 1$.

(b) Compute the sum of the partial quotients of the minimal period of the negative continued fraction expansion of $\omega + 13$, either indirectly from the expansion of $\omega + 12$ or by direct computation of the negative continued fraction (though that requires adding too many partial quotients for our taste; there are eighteen 2s). Confirm that 3 divides the sum.

(c) Deduce the class number $h(-163)$.

Notes

We have seen several proofs that a reduced element is part of a period, or *cycle*, of equivalent reduced elements. Because elements $(\omega + A)/B$ correspond to $\mathbb{Z}[\omega]$-ideals $\langle B, \omega + A \rangle$, we may equally speak of *ideal cycles*. Moreover, a cycle provides a (nontrivial) unit in $\mathbb{Z}[\omega]$; conversely, a unit induces a cycle.

The distance formula entails that the fundamental unit, say $x - \omega y$, provides the *length* $\left|\log |x - \omega y|\right|$ of the cycle. This quantity is also known as the *regulator* of $\mathbb{Z}[\omega]$. Roughly, this length is $\log r$ where r is the number of steps of the period. However, r is usually quite large, $\sqrt{D} \log \log D$ or so. Hence, for serious sizes of D, the units are mostly enormous, typically so big that it is totally unfeasible to display them in any naïve way.

The variation in the period of $\sqrt{n}$ is well illustrated by Figure 4.1, taken from http://oeis.org/A013943/graph.

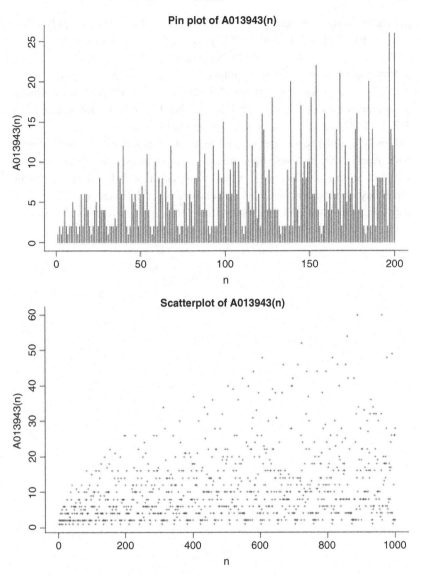

Figure 4.1 The behaviour of the period of the continued fraction of a surd.

In brief, in practice one cannot detail the continuants p_n and q_n. The ideal matrices truly are 'ideal' but only in the sense of 'unreal' or 'theoretical'.

Regarding the final section of this chapter, there are some more wonders of the number 163. The polynomial $f(x) = x^2 + x + 41$ (of discriminant -163)

has the interesting property that $f(0) = 41$, $f(1) = 43$, $f(2) = 47$, $f(3) = 53$, $f(4) = 61$, $f(5) = 71$, $f(6) = 83$, $f(7) = 97$, $f(8) = 113$, $f(9) = 131$, $f(10) = 151, \ldots$, all those values being prime.

Prime numbers p such that the polynomial $f(x) = x^2 + x + p$ is prime for all $x = 1, 2, \ldots, p-1$ are known as *Euler's lucky numbers*. Such primes correspond to imaginary quadratic fields with class number 1. The only such numbers are $2, 3, 5, 11, 17$ and 41 (this is [156, sequence A014556]).

Finally, a well-known practical joke by Martin Gardner (see *Scientific American*, April 1975) involves suggesting to an unsuspecting calculator user that $e^{\pi\sqrt{163}}$ is an integer. This again relies on the fact that 163 corresponds to an imaginary quadratic number field of class number 1 (Exercise 4.25(c)). An amusing related fact is that $\log 640\,320/(3\sqrt{163})$ agrees with π to 15 places.

One more wonder of 163 appears later, in Section 7.3.

5

Hyperelliptic curves and Somos sequences

This chapter, like the previous one and the next, originated in lectures Alf van der Poorten gave in the last few years of his life.

5.1 Two surprising allegations

Consider the following two apparently unrelated mathematical objects.

A pseudo-elliptic integral

$$\int \frac{6X \, dX}{\sqrt{X^4 + 4X^3 - 6X^2 + 4X + 1}}$$
$$= \log\left(X^6 + 12X^5 + 45X^4 + 44X^3 - 33X^2 + 43\right.$$
$$\left. + (X^4 + 10X^3 + 30X^2 + 22X - 11)\sqrt{X^4 + 4X^3 - 6X^2 + 4X + 1}\right). \quad (5.1)$$

REMARK 5.1   Note that pseudo-elliptic integrals as defined below include true *elliptic integrals,* $\int R(w, x) \, dx$, where $R(w, x)$ is a rational function of x and w while w^2 is a cubic or quartic function of x. Unless $R(w, x)$ contains no odd power of w, or w^2 has repeated factors, this is a a non-elementary function of x [28, p. 10].

A Somos sequence of width 5   The sequence

$$\{B_n\}_{n=-\infty}^{\infty} = \{\ldots, 3, 2, 1, 1, 1, 1, 1, 2, 3, 5, 11, 37, 83, \ldots\}$$

is produced by the recursive definition

$$B_{n+3} = (B_{n-1}B_{n+2} + B_nB_{n+1})/B_{n-2}$$

and consists entirely of integers.

Studying the first surprise led van der Poorten to stumble on to the second.

97

The story   More than 20 years ago, Somos noticed that the two-sided sequence $C_{n-2}C_{n+2} = C_{n-1}C_{n+1} + C_n^2$, which we refer to as a 4-*Somos sequence* in his honour, appears to take only integer values if we start from $C_{-1}, C_0, C_1, C_2 = 1$.

Indeed, Somos goes on to investigate also the width-5 sequence

$$B_{n-2}B_{n+3} = B_{n-1}B_{n+2} + B_nB_{n+1},$$

now with five initial 1s, the width-6 sequence

$$D_{n-3}B_{n+3} = D_{n-2}D_{n+2} + D_{n-1}D_{n+1} + D_n^2,$$

and so on, testing whether each – when initiated by an appropriate number of 1s – yields only integers. Naturally, he asked, 'What is going on here?'

Incidentally, while the 4-Somos (A006720 in *Sloane's Online Encyclopedia of Integer Sequences* [156]), 5-Somos (A006721), 6-Somos (A006722) and 7-Somos sequences (A006723) do indeed yield only integers, the analogous 8-Somos sequence does not!

The reality   Somos actually first noticed that the 6-Somos sequence produces only integers while studying θ-function relations; he was subsequently reminded by others that the 4-analogue was known.

Concerning $\{B_n\}_n$ – thus, the 5-Somos sequence – Zagier *inter alia* wrote in [169]:

One computes the first few (in my case, 300) terms B_n numerically, studies their numerical growth, and tries to fit this data by a nice analytic expression. One quickly finds that the growth is roughly exponential in n^2, but with some slow fluctuations around this and also with a dependency on the parity of n. This suggests trying the Ansatz $B_n = c_\pm b^n a^{n^2}$, where $\pm = (-1)^n$. This is easily seen to give a solution to our recursion if a is the root of $a^{12} = a^4 + 1$, and the numerical value $a = 1.07283$ (approx) does indeed give a reasonably good fit to the data, but eventually fails more and more thoroughly. Looking more carefully, we try the same Ansatz but with $c_\pm$ replaced by a function $c_\pm(n)$ which lies between fixed limits but is almost periodic in n, and this works, but with a new value $a = 1.07425$ (approx)...

Here we put in a pause so that the reader can catch his/her breath.

Expanding the function $c_\pm(n)$ numerically into a Fourier series, we discover that it is a Jacobi theta function, and since theta functions (or quotients of them) are elliptic functions, this leads quickly to elliptic curves...

The surprising integral (5.1) is a nice example of a class of *pseudo-elliptic integrals*

$$\int \frac{f(X)\,\mathrm{d}X}{\sqrt{D(X)}} = \log(a(X) + b(X)\sqrt{D(X)}).   \tag{5.2}$$

Here we take D to be a monic polynomial defined over $\mathbb{Q}$, of even degree $2g + 2$ and not the square of a polynomial, and we take f, a and b to denote appropriate polynomials. We suppose a to be nonzero, say of degree $m \geq g + 1$. We will see that necessarily $\deg f = g$, that $\deg b = m - g - 1$ and that f has leading coefficient m. In the example, $m = 6$ and $g = 1$.

Plainly, the equations detailing pseudo-elliptic integrals are purely formal and must therefore remain true when $\sqrt{D}$ is replaced by its conjugate $-\sqrt{D}$. Adding the two conjugate identities we see that

$$\int 0 \, dX = \log(a^2 - Db^2). \tag{5.3}$$

Thus $a^2 - Db^2$ is some constant k and must be nonzero because D is not a square. In other words, $\alpha = a + b\sqrt{D}$ *is a nontrivial unit in the function field* $\mathbb{Q}(X, \sqrt{D(X)})$, and it is immediate that $\deg a = m$ implies $\deg b = m - g - 1$.

Differentiating $\alpha\bar{\alpha}$ yields $2aa' - 2bb'D - b^2 D' = 0$ (the prime here stands for the derivative with respect to X). Hence $b \mid aa'$ and, since a and b must be relatively prime because α is a unit, it follows that $b \mid a'$. Set $f = a'/b$, and note that indeed $\deg f = g$ and that f has leading coefficient m because a and b must have the same leading coefficient. (That common coefficient can be taken as 1, without loss of generality, since we may freely choose the constant produced by the indefinite integration.)

Moreover,

$$\alpha' = a' + b'\sqrt{D} + \frac{bD'}{2\sqrt{D}} = a' + \frac{2bb'D + b^2 D'}{2b\sqrt{D}} = a' + \frac{aa'}{b\sqrt{D}};$$

so, remarkably, we get $\alpha' = f(b\sqrt{D} + a)/\sqrt{D} = f\alpha/\sqrt{D}$.

Thus, to verify the evaluation of the pseudo-elliptic integral (5.2) it suffices to make the *not altogether obvious* substitution $\alpha(x) = a + b\sqrt{D}$. Of course, the real question is how to guess the polynomials $a(x)$ and $b(x)$.

REMARK 5.2 The case $g = 0$, say $D(X) = X^2 + 2uX + w$, is useful for orienting oneself. Here $(X + u) + \sqrt{D(X)}$ is a unit, of norm $u^2 - w$, and indeed

$$\int \frac{dX}{\sqrt{X^2 + 2uX + w}} = \operatorname{arcsinh}\left(\frac{X + u}{\sqrt{w - u^2}}\right) = \log(X + u + \sqrt{X^2 + 2uX + w}).$$

Notice that $\deg f = 0$ and f has leading coefficient 1, as predicted.

5.2 Continued fractions in function fields

In function fields 'polynomial' replaces 'integer' and 'of positive degree' replaces 'positive'. To fix matters we suppose the polynomials to be defined over

some base field K and remark that K may be infinite or finite. A useful analogue for the real numbers is provided by the field of Laurent series $K((X^{-1}))$
of the form

$$F(X) = \sum_{n=-m}^{\infty} f_{-n} X^{-n}.$$

The example series F has degree m and its integer part is defined to be the
polynomial $\lfloor F \rfloor = f_m X^m + f_{m-1} X^{m-1} + \cdots + f_1 X + f_0$.

Matters are exactly as in, or more simple than, the numerical case. Convergents are quotients of relatively prime polynomials and converge to Laurent series (cf. Exercise 2.28); but p/q is a convergent of F iff $\deg(p - Fq) < -\deg q$.
One point that needs care, however, is that the nonzero elements of K are all
(trivial) units of $K[X]$; this fact has some seemingly nontrivial consequences.

Multiplying a continued fraction $[a_0; a_1, a_2, a_3, \ldots]$ by x leads to

$$[xa_0; a_1/x, xa_2, a_3/x, \ldots],$$

with the partial quotients alternately multiplied and divided. An elegant version
of the rule is given by

$$x[ya_0; xa_1, ya_2, xa_3, ya_4, \ldots] = y[xa_0; ya_1, xa_2, ya_3, xa_4, \ldots]. \tag{5.4}$$

Obviously, unless the multiplier is a unit, in general multiplication (or division) leads to drastically inadmissible partial quotients that seriously pollute
the expansion. There are tricks whereby one readies an expansion for multiplication, as in the 'elegant version' above. Alternatively, there is a fine algorithm
of Raney [139], viewing the multiplication as a multiple-state transduction of
an RL-sequence (see Section 4.3).

Even when the multiplication is by a unit, so that no great harm is done,
the effect on the expansion may be startling and unexpected. In the case of
quadratic irrationals over function fields, it creates the possibility of *quasi-
periodicity*, where a 'wannabe' period in fact appears as a sequence of multiples of itself by $k, k^2, k^3, \ldots$

The example $\omega = \sqrt{W^2 + 1}$ gives us trivially

$$\omega + |W| = 2|W| - (\overline{\omega} + |W|),$$

displaying a period length 1.

EXERCISE 5.3

(a) Show that $\mathrm{Norm}(\omega) = \omega\overline{\omega} = -1$.

(b) Is it obvious, or even true, that the example gives all cases of period
    length 1?

It turns out that the correct generalisation of our examples above is the case $\sqrt{W^2 + c}$ with c dividing $4W$. We make the divisibility manifest by considering the case $\sqrt{a^2W^2 + 4a}$.

Suppose we ask much more generally for polynomials $F = F(W)$ such that, as W varies in $\mathbb{Z}$,

(i) $F(W)$ takes only integer values that are not all square and
(ii) the period length of the continued fraction expansion of $\sqrt{|F(W)|}$ is bounded independently of W (thus in terms of F alone).

These questions, specifically (ii), were ingeniously asked and fully answered by Schinzel as early as the 1960s (see [146, 147]). In particular, if F is of odd degree, or if its leading coefficient is not a square, then the periods are certainly unbounded, so we will assume from here on that F has even degree and has square leading coefficient. In this case, the period is bounded if and only if

1. $Y = \sqrt{F(X)}$ has a periodic continued fraction expansion as a quadratic irrational integral function in the domain $\mathbb{Q}[X, Y]$ – such expansions are only periodic by happenstance, because $\mathbb{Q}$ is infinite, and
2. some resulting nontrivial unit of norm dividing 4 in the quadratic function field $\mathbb{Q}(X, Y)$ must have its coefficients in $\mathbb{Z}$, that is, it must be an element of $\mathbb{Z}[X, Y]$.

Patterson and van der Poorten have called this second criterion *Schinzel's condition*. For quadratic F only Schinzel's condition is relevant.

EXERCISE 5.4

(a) Polynomials are usually thought of as having a basis consisting of powers $1, X, X^2, X^3, \ldots$ of the variable. So, if $F(X)$ takes only integer values, it seems natural to guess that all its coefficients must be integers. Not so! Just as good a basis is given by the running powers $1, X, \frac{1}{2}X(X + 1)$, $\frac{1}{6}X(X + 1)(X + 2), \ldots$, so a polynomial of degree s may have denominators as large as $s!$ in its usual presentation and yet take only integer values. Can one do better still?

(b) Show that a polynomial F of even degree and with square leading coefficient may be written uniquely as $F = H^2 + 4R$, where the 'remainder' polynomial $4R$ has degree less than that of the polynomial H.

(c) Hence – this is not at all obvious – show that if F is not the square of a polynomial (equivalently if R is not identically zero) then $F(N)$ cannot be a square for any sufficiently large integer N. It may here be useful to recall that a polynomial of degree s evaluated at N has size of order N^s.

Set $Y^2 = F(W) = a^2W^2 + bW + c$, not a square, and note that Y has polynomial part $aW + b/(2a)$. Then the norm

$$\left(Y + aW + \frac{b}{2a}\right)\left(\overline{Y} + aW + \frac{b}{2a}\right) = \frac{b^2 - 4a^2c}{4a^2}$$

already displays a unit in $\mathcal{R} = \mathbb{Q}[W, Y]$, because any nonzero constant divides 1 in $\mathbb{Q}[W]$.

EXERCISE 5.5

(a) Verify that there is a unit in $\mathcal{R}$ of norm dividing 4 iff $b^2 - 4a^2c$ divides both $16a^4$ and $4b^2$.

(b) Prove that taking $b = 0$ and $c \mid 4a$ yields all the cases above (unless c and a^2 share an odd square factor).

Hint

1. Show that there is no loss of generality in presuming that both a and b are even.

2. Hence make the replacement $b \leftarrow 2b$, and note that the condition becomes $b^2 - a^2c$ divides both $4a^4$ and $4b^2$.

3. Confirm there is now no loss of generality whatsoever in assuming that $0 \le b < |a|$.

4. Show that each case with $b = 0$ corresponds to cases with shorter period than the case $b = 0$.

5. If $b = 0$, deduce there is a unit in $\mathcal{R}$ of norm dividing 4 for all integers W if and only if $c \mid 4a^2$.

6. Show that if p is an odd prime then p times a short period is always at least as long. □

We act on theory and experience by primarily considering an ω given by

(i) $\omega^2 - \omega - \frac{1}{4}(D - 1) = 0$; or

(ii) $\omega^2 - \frac{1}{4}D = 0$, according as $D \equiv 1$ or $0 \pmod 4$;

(iii) $\omega^2 - D = 0$ otherwise.

Thus we obtain the periods of $\sqrt{a^2W^2 - 4c}$ with $c \mid a$, accordingly. Indeed, presuming that $c \mid a$, we have

$$\sqrt{a^2W^2 - 4c} + |aW| = [2|aW|, -\tfrac{1}{2}|aW|/c, \sqrt{a^2W^2 - 4c} + |aW|].$$

So, after division by 2, if aW is odd then

$$\tfrac{1}{2}(1 + \sqrt{a^2W^2 - 4c}) + \tfrac{1}{2}(|aW| - 1) = [\,\overline{|aW|, -|aW|/c}\,]$$

and, when aW is even,

$$\tfrac{1}{2}\sqrt{a^2W^2 - 4c} + \tfrac{1}{2}|aW| = [\,\overline{|aW|, -|aW|/c}\,].$$

In the latter case, since aW is even, this allows us to replace aW by $2aW$ and to obtain

$$\sqrt{a^2W^2 - c} + |aW| = [\,\overline{2|aW|, -2|aW|/c}\,].$$

Therefore if $c \mid a$, and regardless of the parity of aW,

$$\sqrt{a^2W^2 - 2c} + |aW| = [\,\overline{2|aW|, -|aW|/c}\,].$$

If $c \mid a$ but aW is odd, we may multiply by 2 to obtain

$$\sqrt{a^2W^2 - 4c} + |aW| = [\,\overline{2|aW|, -\tfrac{1}{2}(1 + |aW|/c), 2, -\tfrac{1}{2}(1 + |aW|),}$$
$$\overline{2|aW|/c, -\tfrac{1}{2}(1 + |aW|), 2, -\tfrac{1}{2}(1 + |aW|/c)}\,],$$

with a rather longer period than one might naïvely have expected. Confirming this is a nice exercise in multiplying by 2. One indirect way to do that is to use the ideal matrices.

The cases detailed above are intended to be all those for which $c \mid a$ and $a^2W^2 - mc$, with $m = 1$, 2 or 4, is not divisible by a square.

In the foregoing we considered $-c$ rather than c to emphasise the manner in which the sign influences the expansion. Note that we have not observed the requirement that partial quotients should be positive; the purpose of this both to leave an exercise and to make clear that results that appear in the literature as dozens of distinct cases in fact comprise just a handful of cases.

EXERCISE 5.6

(a) Above we speak of 'dozens of different cases'. If these computations were completed by rewriting each expansion so that it has only positive partial quotients, and with c both positive and negative, how many different cases do in fact result?

(b) Rewrite several of the cases.

(c) (For 'negative' readers.) Redo (a) and (b) so as to obtain partial quotients with alternating sign.

5.3 Units and torsion

Recall that the pseudo-elliptic integral evaluation (5.2) results in the logarithm of a unit α in the function field $\mathbb{Q}[X, \sqrt{D(X)}]$. The notion 'unit' entails that α

be trivial at other than infinite places (i.e. absolute values). That is, the divisor of the zeros and poles of the function $\alpha = a + b\sqrt{D}$ is supported only at infinity.

Plainly speaking, the quartic $C : Y^2 = X^4 + 4X^3 - 6X^2 + 4X + 1$ corresponding to the original integral (5.1) has two points at infinity, which we shall call S and O – the latter being the zero of the group law on the elliptic curve C. In general, for $C : Y^2 = D(X)$ of genus g, we are unfortunately forced to talk about the divisor at infinity on C or, worse, about its class as a point on the Jacobian of the hyperelliptic curve C. The hyperelliptic curve $y^2 = x^5 - 5x^3 + 4x$ is illustrated in Figure 5.1.

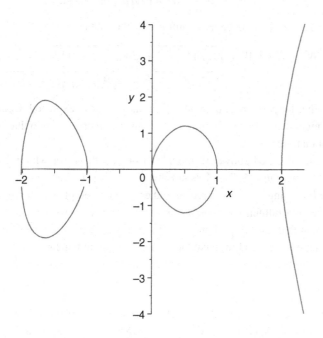

Figure 5.1 The hyperelliptic curve $y^2 = x^5 - 5x^3 + 4x$.

Whenever we meet a pseudo-elliptic integral, there is a positive integer m such that $m(S - O)$ is the divisor of a unit α, showing that $S - O$ provides a *torsion point* of order m on Jac(C), the Jacobian of C.

One finds a unit α in the quadratic domain $\mathbb{Q}[X, Y]$ by studying the continued fraction expansion of $Y = \sqrt{D(X)}$. The principle is that a period of the expansion produces a unit and, conversely, the existence of a unit entails the periodicity of the continued fraction expansion (see Theorem 4.14).

Thus – because periodicity is equivalent to torsion at infinity – *each step in the continued fraction expansion of Y must somehow add some multiple of the divisor at infinity.* This fact was nicely made explicit in the elliptic case by Adams and Razar [3]; there is a more recent analogous proof for general g by Berry [14].

It's pretty obvious that torsion at infinity is unusual in characteristic zero. So, the periodicity of the expansion of Y must therefore be exceptional.

In the numerical case, and for congruence function fields, periodicity is always forced by the box principle. But, over an infinite field, there are infinitely many polynomials of bounded degree; so periodicity is a rare happenstance.

Set $Y^2 = D(X)$ where D is a nonsquare monic polynomial of degree $2g + 2$. Then (see Exercise 5.4) we may write $D(X) = H(X)^2 + 4R(X)$, where H is the polynomial part of the square root Y of D, and $4R$, with $\deg R \leq g$, is the remainder. We then take

$$Y = H\left(1 + \frac{4R}{H^2}\right)^{1/2} = H(X) + c_1 X^{-1} + c_2 X^{-2} + \cdots ,$$

thereby viewing Y as an element of $K((X^{-1}))$, the Laurent series in the variable $1/X$. All this makes sense over any base field K not of characteristic 2.

However, below we deal with the quadratic irrational function Z defined by

$$C : Z^2 - HZ - R = 0; \qquad \text{in effect} \quad Z = \tfrac{1}{2}(Y + H). \tag{5.5}$$

Then $\deg Z = \deg H = g + 1$, while its conjugate satisfies $\deg \overline{Z} < 0$; so Z is reduced. Note that Z makes sense for arbitrary characteristic, including characteristic 2.

Let $Z^2 - HZ - R = 0$. In what follows Z will denote a nontrivial quadratic irrational function of polynomial trace H and polynomial norm $-R$, with $\deg R < \deg H$. The word 'irrational' entails that Z is not a polynomial; thus $R \neq 0$.

EXERCISE 5.7   Confirm that:

(a) given that Z is in $K((X^{-1}))$, there plainly is no loss of generality in supposing, as we have, that $\deg R < \deg H$ or equivalently that H is the polynomial part of $\sqrt{D} = Z - \overline{Z}$, the square root of the discriminant $D = H^2 + 4R$ of Z; and

(b) given that $\deg Z > \deg \overline{Z}$, the conditions $\deg Z > 0$ and $\deg \overline{Z} < 0$ precisely affirm that Z is *reduced*, in the sense that the continued fraction process on a quadratic irrational always leads to, and then sustains, the conditions.

Technically, Z is a real quadratic irrational function; quadratic irrationals defined over $K[X]$ but not in $K((X^{-1}))$ are imaginary.

To detail the continued fraction expansion, for $n \in \mathbb{Z}$ set

$$Z_n = (Z + A_n)B_n,$$

where A_n and B_n are polynomials with B_n dividing the norm $(Z + A_n)(\overline{Z} + A_n)$. Suppose that $\deg Z_n > 0$ and $\deg \overline{Z}_n < 0$, in other words that Z_n is reduced.

EXERCISE 5.8

(a) Show that Z_n is reduced iff $\deg A_n \leq g - 1$ and $\deg B_n \leq g$.
(b) Let a_n denote the polynomial part of Z_n, and set $Z_n = a_n - \overline{R}_{-n}$. Imitate the argument of the numerical case to confirm that R_{-n} and $Z_{n+1} = -1/\overline{R}_{-n}$ are reduced.

It seems to follow that every reduced element must have a purely periodic continued fraction expansion. And that's true, but only in a sense. The trouble is that if the base field K is infinite then the period is generically of infinite length – the box principle does not apply because if K is infinite then there are infinitely many polynomials of bounded degree.

More, it is then a rare and unusual happenstance for any reduced Z_0 to have a periodic expansion.

Now insist that $\deg Z_0 > 0$ and $\deg \overline{Z}_0 < 0$. Then $\deg Z_n > 0$ and $\deg \overline{Z}_n < 0$, that is, the Z_n all are *reduced*, and the $K[X]$-module

$$\langle B_n, Z + A_n \rangle$$

is in fact an ideal of the domain $K[X, Z]$.

Finally, let a_n denote the polynomial part of Z_n. Then the continued fraction expansion of, say, Z_0 is a table of lines (or steps)

$$\frac{Z + A_n}{B_n} = a_n - \frac{\overline{Z} + A_{n+1}}{B_n};$$

in brief, $Z_n = a_n - \overline{R}_{-n}$ where

$$-\frac{B_n}{\overline{Z} + A_{n+1}} = \frac{Z + A_{n+1}}{B_{n+1}}.$$

Necessarily,

$$A_n + A_{n+1} + H = a_n B_n \qquad \text{and} \qquad (Z + A_{n+1})(\overline{Z} + A_{n+1}) = -B_n B_{n+1},$$

and one easily verifies that the conditions on the A_n and B_n are fulfilled.

There is a minor miracle: because all the complete quotients Z_n are reduced, it follows that all the 'remainders' R_{-n} are reduced, too. Thus, the partial quotients a_n, which began life as the polynomial parts of the Z_n, also are the polynomial parts of the R_{-n}.

Furthermore, the conjugate line

$$R_{-n} = \frac{Z + A_{n+1}}{B_n} = a_n - \frac{\overline{Z} + A_n}{B_n} = a_n - \overline{Z}_n$$

is a line in an admissible continued fraction expansion, which explains why we can refer to the original expansion as *bidirectionally* infinite.

When the base field K is infinite (say $K = \mathbb{Q}$), we assert that a generic choice of A_0 and B_0 is such that all the a_n are linear – equivalently, such that all the B_n are of degree g or, indeed, a little bit less obviously, such that all the A_n are of their maximal degree, $g - 1$. That's so because the probability of an element of K being zero is zero. It is equivalent to note that a generic divisor of C is defined by a g-tuple of elements of an algebraic extension of K.

We should point out that any actual expansion is very messy. We will give a list of the partial quotients for two very different examples. First,

$$\sqrt{X^4 - 2X^3 + 3X^2 + 2X + 1} + (X^2 - X + 1)$$

$$= \overline{\left[\, 2(X^2 - X + 1), \tfrac{1}{2}(X - 1), 2(X - 1), \tfrac{1}{2}(X^2 - X + 1), 2(X - 1), \tfrac{1}{2}(X - 1)\,\right]}.$$

Here, we have copied the expansion of $2Z$ in a periodic case (so there is a pseudo-elliptic integral with $D = X^4 - 2X^3 + 3X^2 + 2X + 1$). Note that the quasiperiod already supplies a unit. In fact,

$$\int \frac{(4X - 1)\,\mathrm{d}X}{\sqrt{X^4 - 2X^3 + 3X^2 + 2X + 1}}$$

$$= \log(X^4 - 3X^3 + 5X^2 - 2X + (X^2 - 2X + 2)\sqrt{X^4 - 2X^3 + 3X^2 + 2X + 1}).$$

Second, if we replace D by $D + 1$ then we obtain a generic expansion nicely illustrating the behaviour of the *Néron–Tate height*:

$$\sqrt{X^4 - 2X^3 + 3X^2 + 2X + 2} + (X^2 - X + 1)$$

$$= \Big[\, 2(X^2 - X + 1), \tfrac{1}{2}X - \tfrac{5}{8}, \tfrac{32}{21}X - \tfrac{344}{441}, -\tfrac{9\,261}{7\,936}X - \tfrac{2\,963\,079}{1\,968\,128},$$

$$-\tfrac{488\,095\,744}{2\,572\,789\,149}X + \tfrac{16\,216\,931\,891\,200}{34\,035\,427\,652\,121},$$

$$-\tfrac{21\,440\,698\,686\,186\,129}{1\,136\,033\,082\,245\,120}X + \tfrac{1\,665\,322\,334\,299\,891\,329\,867}{42\,646\,681\,907\,481\,804\,800},$$

$$-\tfrac{1\,600\,956\,438\,806\,866\,952\,192\,000}{88\,607\,770\,352\,600\,487\,715\,818\,861}X - \tfrac{3\,371\,996\,766\,956\,576\,002\,150\,497\,085\,030\,400}{256\,351\,315\,939\,101\,539\,512\,201\,711\,796\,263\,641},$$

$$\tfrac{80\,083\,198\,356\,049\,188\,999\,341\,382\,795\,525\,473\,293\,961}{975\,968\,207\,083\,235\,989\,098\,500\,587\,163\,484\,160\,000}X$$

$$-\tfrac{255\,369\,300\,674\,062\,782\,420\,731\,816\,474\,523\,944\,637\,364\,177\,546\,099}{12\,679\,074\,228\,671\,726\,095\,323\,776\,878\,469\,612\,834\,847\,195\,136\,000}, \cdots \Big].$$

Even a digital computer chokes on numbers growing at such a pace.

In the course of studying the continued fraction expansions

$$\frac{Z + A_n}{B_n} = a_n - \frac{\overline{Z} + A_{n+1}}{B_n}, \qquad n \in \mathbb{Z},$$

in quadratic function fields one can learn that the leading coefficients, say d_n, of the polynomials A_n control the degeneracies of the expansion. So, we will now choose to describe the other parameters detailing the A_n and B_n in terms of the d_n.

Let ϑ_n denote a typical zero of B_n, and recall the recursion relations

$$A_n + A_{n+1} + H = a_n B_n$$

and

$$-B_n B_{n+1} = (Z + A_{n+1})(\overline{Z} + A_{n+1}) = -R + A_{n+1}(H + A_{n+1}).$$

Then

$$A_n(\vartheta_n) + A_{n+1}(\vartheta_n) + H(\vartheta_n) = 0 \qquad \text{and} \qquad R(\vartheta_n) = -A_{n+1}(\vartheta_n)A_n(\vartheta_n).$$

Hence $B_n(X)$ divides $R(X) + A_{n+1}(X)A_n(X)$, and so

$$\frac{C_n(X)}{s_n} = \frac{R(X) + A_{n+1}(X)A_n(X)}{B_n(X)}$$

defines a polynomial C_n; here s_n is the leading coefficient of B_n. It is useful that $\deg C_n = \max\{g, 2(g-1)\} - g$, so C_n *is a constant* if $g = 1$ or $g = 2$.

Note that $(Z + A_n)(\overline{Z} + A_n) = -R + HA_n + A_n^2$. Suppose ϑ_n denotes a typical zero of B_n. Then the condition 'B_n divides the norm $(Z + A_n)(\overline{Z} + A_n)$' asserts that $R(\vartheta_n) = (H(\vartheta_n) + A_n(\vartheta_n))A_n(\vartheta_n)$.

However, recall that Z defines the hyperelliptic curve

$$C : Z^2 - HZ - R = 0$$

of *genus g*. Of course, C is defined over K but, for a moment disregarding that fact, it follows from the remarks above that the point $(\vartheta_n, -A_n(\vartheta_n))$ lies on C. In general $\deg B_n = g$ and so has g conjugate zeros. That gives a g-tuple of conjugate points on C or, in the proper language, a *divisor* defined over K on C.

As we have already seen, the consecutive divisor classes differ by some multiple (in fact the degree of a_n) of the class of the divisor S at infinity on (the Jacobian of) C. If $\deg a_n = 1$ for all n then this provides a sequence $\{M_n\}_n$ of divisors, with $M_n = M + nS$.

If $g = 1$, C is an elliptic curve equal to its Jacobian and this story is about honest-to-goodness points on the curve.

5.4 The elliptic case

When $g = 1$ we have $\deg A_n = 0$; we will set $A_n = d_n$ and $\deg B_n = 1$, say with $B_n(X) = s_n(X - t_n)$. Then $\deg H = 2$. Now set, say, $R = s(X - t)$.

Happily, the birational transformation $U = Z$, $V - s = XZ$, transforms our quartic curve into a cubic model passing through the origin:

$$\mathcal{E} : V^2 - sV = \text{monic cubic in } U \text{ with zero constant term};$$

the points $(t_n, -d_n)$ on C become $(-d_n, s - d_n t_n)$ on $\mathcal{E}$. The point S is now $(0, 0)$. It is then easy to use the continued fraction recursion formulae to verify explicitly that $S_n = nS$.

We have that $-R(t_n) = d_n d_{n+1}$, that $C_n = s$ and that $-C_{n-1} C_n B_{n-1}(t) B_n(t)$ equals both $s^2 A_n(t)(H(t) + A_n(t))$ and $-s_{n-1} s_n d_{n-1} d_n^2 d_{n+1}$. Thus

$$d_{n-1} d_n^2 d_{n+1} = s^2(d_n + H(t)), \qquad (5.6)$$

a recursion involving the d_n only. But the d_n are very messy!

The $-d_n$ are in fact the abscissae of points on $\mathcal{E}$ (specifically, of the points $M + nS$); so they are rationals whose denominators, D_n^2, say, are squares of integers. Accordingly, define a sequence $\{D_n\}_n$ by

$$D_{n-1} D_{n+1} = d_n D_n^2.$$

Conveniently, this yields immediately

$$D_{n-2} D_{n+2} = d_{n-1} d_n^2 d_{n+1} D_n^2,$$

so (5.6) becomes

$$D_{n-2} D_{n+2} = s^2 D_{n-1} D_{n+1} + s^2 H(t) D_n^2,$$

showing that all 4-Somos sequences come from (at most quadratic twists of) rational elliptic curves.

A careful look (see, for example, Shipsey's thesis [152]) at the behaviour of points $M + nS$ on an elliptic curve confirms that all the D_n are S-integers, the primes of the finite set S coming from the factors of the initial values D_0, D_1, D_2, D_3 and the denominators of the coefficients s and $H(t)$ of the Somos relation.

As it happens, a combinatorial algebraic result, a corollary of Fomin's and Zelevinsky's theory of cluster algebras [59], guarantees that elements of 4-, 5-, 6- and 7-Somos sequences are Laurent polynomials in the initial values with coefficient ring polynomials in the coefficients of the defining recursion.

Sequences that are 5-Somos also come from elliptic curves. It is not difficult to find that

$$D_{n-1} D_{n+2} = d_n d_{n+1} D_n D_{n+1}.$$

The ingenious observation that

$$d_{n+1}d_n + \frac{s^2}{d_n} + d_n d_{n-1}$$

is independent of n – that is, it is a discrete integral of the recursion for the d_n –
now readily yields an identity providing the width-5 recursion

$$D_{n-2}D_{n+3} = -s^2 H(t) D_{n-1} D_{n+2} + s^3 (s + 2tH(t)) D_n D_{n+1}.$$

A 5-Somos sequence may also be a 4-Somos sequence. In any case, its two
subsequences $\{D_{2n+1}\}_n$ and $\{D_{2n}\}_n$ are different 4-Somos sequences deriving
from just one elliptic curve, with the same addition by $S_\mathcal{E} = (0,0)$, but with
initial translations $M_\mathcal{E}$ differing by $\frac{1}{2}S_\mathcal{E}$.

Now consider the singular case, where $M = S$, and the expansion of Z. It
will be convenient to write e in place of d, and – in honour of Ward – $\{W_n\}_n$ in
place of $\{D_n\}_n$. A brief computation reveals that $a_0(X) = H$, $e_1 = 0$, $B_1(X) = s(X - t)$ and $e_2 = -H(t)$; it then suffices, using the recursion for the sequence
$\{d_n\}_n$, to set $W_1 = 1$, $W_2 = s$, leading to $W_3 = -s^2 H(t)$, $W_4 = -s^4(s + 2tH(t))$
and so on. One then notices that $\{W_n\}_n$ supplies the coefficients in

$$D_{n-2}D_{n+2} = W_2^2 A_{n-1} A_{n+1} - W_1 W_3 A_n^2.$$

Van der Poorten realised much later that this was obvious and did not need
any work. The point is that the recursion leading to the D_n is independent of
the translation M. So, the coefficients above must be precisely those that fit the
singular case $D_n = W_n$. Van der Poorten twigged this self-determining principle after Stange first showed him her work on elliptic nets; he then realised
that he had already used this principle in work with Swart [136].

Remarkably [165], Ward introduced his sequence $\{W_n\}_n$ as the one satisfying
$W_{-n} = -W_n$ and the *multi-recursion*

$$W_k^2 W_{n-m} W_{n+m} = W_m^2 W_{n-k} W_{n+k} - W_{m-k} W_{m+k} W_n^2.$$

However, the special case $k = 1$, $m = 2$ and the values W_1, W_2, W_3, W_4 already
determine the sequence.

Ward proved the coherence of his definition by showing that there does exist a solution sequence defined in terms of Weierstrass σ-functions. Van der
Poorten and Swart re-explored this matter [136] and found a direct proof that
for all k and m

$$W_k^2 D_{n-m} D_{n+m} = W_m^2 D_{n-k} D_{n+k} - W_{m-k} W_{m+k} D_n^2.$$

Their argument relies on the amusingly symmetrical identity

$$(d_{n-1} - e_m) d_n^2 (d_{n+1} - e_m) = (e_{m-1} - d_n) e_m^2 (e_{m+1} - d_n).$$

They had a similar argument and an analogous result for the odd-gap case.

We insisted that the cubic model $\mathcal{E}$ of our elliptic curve should contain $(0, 0)$. However, we may suppose that we have obtained our $\mathcal{E} = \mathcal{E}(x, y)$ from a more general elliptic curve by translating a point $S = (x, y)$ on it to the origin. Then the coefficients of $\mathcal{E}$ are polynomials in x and y, themselves having coefficients polynomial in the defining data. This makes the W_n polynomials in x and y.

Moreover, $mS = 0$ and $W_m(a, b) = 0$ iff $S = (a, b)$ is a torsion point of order m on $\mathcal{E}$. So the polynomial $W_n(x, y)$ is the nth division polynomial. This *inter alia* entails that

$$\gcd(W_m(x, y), W_n(x, y)) = W_{\gcd(m,n)}(x, y),$$

explaining the division properties of the $W_n(0, 0)$.

Incidently, Shipsey [152] proved directly that if $W_1 = 1$ and $W_2 \mid W_4$ then $m \mid n$ entails $W_m \mid W_n$; hence again we have an elliptic *divisibility* sequence.

4-*Somos sequences* Suppose that $\{C_n\}_n = \{\ldots, 2, 1, 1, 1, 1, 2, 3, 7, \ldots\}$ with $C_{n-2}C_{n+2} = C_{n-1}C_{n+1} + C_n^2$. Our formula quickly reveals that $s = \pm 1$, $t = \mp 2$, $H(t) = 1$, and thus the sequence $\{C_n\}_n$ arises from

$$Z^2 - (X^2 - 3)Z - (X - 2) = 0 \qquad \text{with } M = (1, -1)$$

or, equivalently, from

$$\mathcal{E} : V^2 - V = U^3 + 3U^2 + 2U \qquad \text{with } M_\mathcal{E} = (1, -1).$$

5-*Somos sequences* The case $\{B_n\}_n = \{\ldots, 2, 1, 1, 1, 1, 1, 2, 3, 5, 11, \ldots\}$ with

$$B_{n-2}B_{n+3} = B_{n-1}B_{n+2} + B_nB_{n+1}$$

is trickier. One needs to define $c_nB_{n-1}B_{n+1} = e_nB_n^2$ with c_nc_{n+1} independent of n. It turns out that $\{B_n\}_n$ arises from

$$Z^2 - (X^2 - 29)Z - 48(X + 5) = 0 \qquad \text{with } M = (-3, -10)$$

or, equivalently, from

$$\mathcal{E} : V^2 + UV + 6V = U^3 + 7U^2 + 12U \qquad \text{with } M_\mathcal{E} = (-2, -2).$$

The fact that $\gcd(6, 12) \neq 1$ prevents the sequence from being a 4-Somos sequence.

By symmetry each respective M is a point of order 2 on the corresponding curve.

5.5 Higher genus

There certainly are analogous results for higher-genus curves, particularly for hyperelliptic curves. Indeed, Cantor showed [41] for all hyperelliptic curves of odd degree that there are analogues of the division polynomials satisfying relations given by certain Kronecker–Hankel determinants.

A continued fraction program usually falters almost immediately, though it can completely handle curves $Z^2 - HZ - R = 0$ with $\deg H = 3$ provided that $R(X) = -s(X - t)$ is linear. In that case, we find that

$$d_{n-2}d_{n-1}^2 d_n^3 d_{n+1}^2 d_{n+2} = s^2 d_{n-1}d_n^2 d_{n+1} - s^3 H(t),$$

yielding a width-6 relation

$$D_{n-3}D_{n+3} = s^2 D_{n-2}D_{n+2} - v^3 H(t)D_n^2.$$

Others have done worse, and better. If $g = 1$, Hone noted [75] that $\alpha = \wp'(y)^2, \beta = \wp'(y)^2(\wp(2y) - \wp(y))$ yields the identity

$$(\wp(x+y) - \wp(y))(\wp(x) - \wp(y))^2(\wp(x+y) - \wp(y)) = -\alpha(\wp(x) - \wp(y)) + \beta.$$

This was also a remark of Stephens basic to Swart's thesis [158]. Notice that it is precisely van der Poorten's relation on the $-d_n$.

Hone *et al.* found an analogous relation for Kleinian σ-functions in genus 2 and used it to obtain a 8-Somos (not the most general 8-Somos) relation corresponding to curves given by $Y^2 = $ a quintic in X.

Van der Poorten knew, in part on the basis of results of Cantor, that for $g = 2$ the minimal relation has in fact width 6 but is cubic, rather than quadratic as in the Somos cases.

Surprisingly, perhaps, this does cohere with a suggestion of Elkies that the special cases $Z^2 - HZ + s(X - t) = 0$ with $\deg H = g + 1$ do yield Somos relations of width $2g + 2$.

A cute example à la Somos In any case, we can apply continued fraction methods to find that $\{T_n\}_n = \{\ldots, 2, 1, 1, 1, 1, 1, 1, 2, 3, 4, 8, 17, 50, \ldots\}$, with

$$T_{n-3}T_{n+3} = T_{n-2}T_{n+2} + T_n^2,$$

arises from the points (thus, divisor classes) $\ldots, M - S, M, M + S, M + 2S, \ldots$ on the Jacobian of the genus-2 hyperelliptic curve

$$C : Z^2 - (X^3 - 4X + 1)Z - (X - 2) = 0.$$

Here S is the class of the divisor at infinity and M is instanced by the divisor defined by the pair of points $(\varphi, \overline{\varphi})$ and $(\overline{\varphi}, \varphi)$, where φ is the golden ratio.

The symmetry dictates that $M - S = -M$, so that $2M = S$ on Jac(C).

Notes

Continuing our citation of Zagier in [169]:

By the way, by varying the coefficients in the recursion for the B_n one can replace the above [elliptic curve with the rational point on it] by essentially any elliptic curve over $\mathbb{Q}$ and any rational point on it, so that in an elementary course in number theory one could develop (or at least introduce) the entire theory of elliptic curves just by starting with these simple recursions!

Rather than attempt a listing here, we refer the reader to the arXiv e-print archive (http://arxiv.org/), where a search with keyword 'Somos' will find many papers that we have cited implicitly. The references of those papers will readily lead to a fairly full list of the relevant literature.

6

From folding to Fibonacci

This is the third of the three chapters based on lectures of Alf van der Poorten. What two activities could seem less related than continued fraction expansion and the folding of a piece of paper – see Figure 6.1? Nevertheless, we will see in this chapter that neverending folding is connected to neverending fractions.

6.1 Folding

Take a piece of paper and fold it once, then unfold to 90°. You get something like this: ∧. Now do the same thing but fold it twice, once over itself. This time, unfolding it produces a pattern like this: ∧ ∧ ∨. Repeating the folding n times produces a sequence of $2^n - 1$ 'hills' (which we represent by 1) and 'valleys' (which we represent by $-1 = \bar{1}$). Here are the first few sequences, where ϵ denotes the empty string of length 0 (see also Figure 6.1):

number of folds	pattern
0	ϵ
1	1
2	$1\,1\,\bar{1}$
3	$1\,1\,\bar{1}\,1\,1\,\bar{1}\,\bar{1}$
4	$1\,1\,\bar{1}\,1\,1\,\bar{1}\,\bar{1}\,1\,1\,1\,\bar{1}\,\bar{1}\,1\,\bar{1}\,\bar{1}$

We can describe the resulting pattern using the *folding map* F_a, which sends w, a string of integers, to $(w, a, -^t w)$; here, as in Section 2.14, the left upper index t reverses a string, a minus sign in front negates each entry and the commas denote concatenation.

The folding we have been doing so far corresponds to F_1, but one could just as well choose a different kind of folding, where we insert a valley after a fold,

114

Figure 6.1 Paper folded three times and then unfolded as described in Section 6.1.

instead of a hill – the difference between folding right-hand over left and vice versa. At each step, then, we can choose between performing F_1 or $F_{\bar{1}}$.

Observation 6.1 *Suppose that we fold with instructions $a_1, a_2, \ldots, a_n$ and then unfold all to 90°. The resulting sequence of folds is given by*

$$F_{a_1}(F_{a_2}(\cdots F_{a_n}(\epsilon)\cdots)).$$

Proof First, fold the paper once with a_1, but don't tell the paper it has been folded, that is, the paper has no memory of being folded. Now fold it in this way $n - 1$ further times, obtaining (by induction)

$$x = F_{a_2}(F_{a_3}(\cdots F_{a_n}(\epsilon)\cdots))$$

on the once-folded paper. Finally, do the last unfolding. The process of doing this unfolding reveals a_1 as the inserted fold, and 'swinging' out the paper for the last unfolding both reverses the order of the folds and changes hills into valleys (and vice versa). Thus $F_{a_1}(x) = (x, a_1, {}^{-t}x)$. □

So far we have folded finitely. To discuss an infinite folding it is conceptually easier to reverse the order of the folds, so that the finite list of folding instructions $a_1, a_2, \ldots, a_n$, when reversed, becomes the list of *unfolding instructions* $b_0 = a_n, b_1 = a_{n-1}, \ldots, b_{n-1} = a_1$. We define

$$\mathrm{Unf}(b_0, b_1, \ldots, b_n) = F_{a_1}(F_{a_2}(\cdots F_{a_n}(\epsilon)\cdots)).$$

As $n \to \infty$, this associates each infinite sequence of unfolding instructions $b = (b_i)_{i \geq 0}$ with a unique infinite sequence of folds.

EXERCISE 6.2 Show that if $(f_i)_{i \geq 1} = \mathrm{Unf}(b_0, b_1, \ldots)$ then $f_n = (-1)^j b_k$, where j, k are the unique nonnegative integers such that $n = 2^k(2j + 1)$.

Now it's time to turn to continued fractions. The main tool is the so-called *folding lemma*:

Lemma 6.3 *Let $w = (a_1, a_2, \ldots, a_n)$ be a string of partial quotients, and let $p_n/q_n = [a_0; w]$. Then*

$$\frac{p_n}{q_n} + \frac{(-1)^n}{tq_n^2} = [a_0; w, t, -{}^t w].$$

Proof We use the 2×2 matrix approach of Lemma 2.8. There it was proved that

$$\begin{pmatrix} a_0 & 1 \\ 1 & 0 \end{pmatrix} \begin{pmatrix} a_1 & 1 \\ 1 & 0 \end{pmatrix} \cdots \begin{pmatrix} a_n & 1 \\ 1 & 0 \end{pmatrix} = \begin{pmatrix} p_n & p_{n-1} \\ q_n & q_{n-1} \end{pmatrix}.$$

In a similar fashion, we can prove that

$$\begin{pmatrix} -a_0 & 1 \\ 1 & 0 \end{pmatrix} \begin{pmatrix} -a_1 & 1 \\ 1 & 0 \end{pmatrix} \cdots \begin{pmatrix} -a_n & 1 \\ 1 & 0 \end{pmatrix} = \begin{pmatrix} (-1)^{n+1} p_n & (-1)^n p_{n-1} \\ (-1)^n q_n & (-1)^{n-1} q_{n-1} \end{pmatrix}.$$

Take the transpose of both sides and multiply by

$$\begin{pmatrix} -a_0 & 1 \\ 1 & 0 \end{pmatrix}^{-1} = \begin{pmatrix} 0 & 1 \\ 1 & a_0 \end{pmatrix}$$

to get

$$\begin{pmatrix} -a_n & 1 \\ 1 & 0 \end{pmatrix} \begin{pmatrix} -a_{n-1} & 1 \\ 1 & 0 \end{pmatrix} \cdots \begin{pmatrix} -a_1 & 1 \\ 1 & 0 \end{pmatrix} = \begin{pmatrix} (-1)^n q_n & * \\ (-1)^{n-1} q_{n-1} & * \end{pmatrix},$$

where an asterisk represents an entry we don't care about. Putting this all together, we get

$$\begin{pmatrix} a_0 & 1 \\ 1 & 0 \end{pmatrix} \cdots \begin{pmatrix} a_n & 1 \\ 1 & 0 \end{pmatrix} \begin{pmatrix} t & 1 \\ 1 & 0 \end{pmatrix} \begin{pmatrix} -a_n & 1 \\ 1 & 0 \end{pmatrix} \cdots \begin{pmatrix} -a_1 & 1 \\ 1 & 0 \end{pmatrix}$$

$$= \begin{pmatrix} (-1)^n (q_n(tp_n + p_{n-1}) - p_n q_{n-1}) & * \\ (-1)^n (q_n(tq_n + q_{n-1}) - q_n q_{n-1}) & * \end{pmatrix}.$$

So, we obtain as desired

$$[a_0; w, t, -{}^t w] = \frac{q_n(tp_n + p_{n-1}) - p_n q_{n-1}}{q_n(tq_n + q_{n-1}) - q_n q_{n-1}}$$

$$= \frac{p_n}{q_n} + \frac{(-1)^n}{tq_n^2}. \qquad \square$$

Finally, let's apply all this to find the continued fraction expansion of the formal Laurent series $x \sum_{n=0}^{\infty} x^{-2^n} = 1 + x^{-1} + x^{-3} + x^{-7} + \cdots$:

Corollary 6.4 *We have*

$$x \sum_{n=0}^{\infty} x^{-2^n} = [1; \mathrm{Unf}(x, -x, -x, -x, -x, \ldots)]$$

$$= [1; x, -x, -x, -x, x, x, -x, -x, x, -x, -x, x, x, x, -x, \ldots].$$

Proof Start with $[1, x]$ and successively set $t = -x$ in the folding lemma to get the result. □

EXERCISE 6.5 Specialise the expansion in Corollary 6.4 by setting $x = 2$. Then remove the inadmissible partial quotients to show that the real number $2 \sum_{n=0}^{\infty} 2^{-2^n}$ has a continued fraction expansion in which the partial quotients are just 1s and 2s.

6.2 Zaremba's conjecture

A beautiful and still unresolved conjecture due to Zaremba [172], mentioned in the end notes for Chapter 3, involves the sizes of partial quotients. He conjectured the existence of an absolute constant C with the following property: for all integers $m \geq 1$, there exists a positive integer a, relatively prime to m, such that all the partial quotients in the continued fraction expansion for a/m are bounded by C. Indeed, it is widely believed that $C = 5$ suffices, and even $C = 3$ for $m \geq 6235$.

Here we show how to use the folding lemma to prove Zaremba's conjecture when m is a power of 2. Similar ideas can be used to prove it for powers of 3, 5 and 6.

Theorem 6.6 *For all integers $n \geq 0$ there exists an odd positive integer a such that the continued fraction for $a/2^n$ has partial quotients bounded by 3.*

Proof For $1 \leq n \leq 5$ we have

$$1/2 = [0; 2],$$
$$3/4 = [0; 1, 3],$$
$$3/8 = [0; 2, 1, 2],$$
$$7/16 = [0; 2, 3, 2],$$
$$25/32 = [0; 1, 3, 1, 1, 3].$$

For $n \geq 6$ we assert a stronger claim, which we prove by induction: namely, that there exists a such that $a/2^n$ has a continued fraction expansion of the form $[0; 2, \ldots, 2]$ with all the elided partial quotients ≤ 3.

For $6 \leq n \leq 11$ this follows from

$$23/64 = [0; 2, 1, 3, 1, 1, 2],$$
$$47/128 = [0; 2, 1, 2, 1, 1, 1, 1, 2],$$
$$95/256 = [0; 2, 1, 2, 3, 1, 1, 1, 2],$$
$$223/512 = [0; 2, 3, 2, 1, 1, 1, 3, 2],$$
$$367/1024 = [0; 2, 1, 3, 1, 3, 3, 1, 1, 2],$$
$$791/2048 = [0; 2, 1, 1, 2, 3, 3, 1, 1, 2, 2].$$

For $n \geq 12$ we proceed by induction. Suppose n is odd, so that $n = 2m + 1$. Given an expansion $a/2^m = [0; 2, a_2, \ldots, a_{s-1}, 2]$, we use the folding lemma with $t = 2$ to find

$$\frac{a \times 2^{m+1} + 1}{2^{2m+1}} = \frac{a}{2^m} + \frac{1}{2 \times 2^{2m}}$$
$$= [0; 2, a_2, \ldots, a_{s-1}, 2, 2, -2, -a_{s-1}, \ldots, -a_2, -2]$$
$$= [0; 2, a_2, \ldots, a_{s-1}, 2, 2, 0, -1, 1, -1, 0, 2, a_{s-1}, \ldots, a_2, 2]$$
$$= [0; 2, a_2, \ldots, a_{s-1}, 2, 1, 1, 1, a_{s-1}, \ldots, a_2, 2],$$

which maintains the invariant. Here we have used the negation lemma of Example 4.21. Now suppose n is even, say $n = 2m$. From the folding lemma with $t = 1$ we get

$$\frac{a \times 2^m + 1}{2^{2m}} = \frac{a}{2^m} + \frac{1}{2^{2m}}$$
$$= [0; 2, a_2, \ldots, a_{s-1}, 2, 1, -2, -a_{s-1}, \ldots, -a_2, -2]$$
$$= [0; 2, a_2, \ldots, a_{s-1}, 2, 1, 0, -1, 1, -1, 0, 2, a_{s-1}, \ldots, a_2, 2]$$
$$= [0; 2, a_2, \ldots, a_{s-1}, 3, 1, a_{s-1}, \ldots, a_2, 2],$$

again maintaining the invariant. □

EXERCISE 6.7 Prove Zaremba's conjecture (with constant $C = 3$) when the denominators are powers of 3.

6.3 On to Fibonacci

Many years ago now, one of us noticed the continued fraction expansion

$$2^{-1} + 2^{-2} + 2^{-3} + 2^{-5} + \cdots + 2^{-F_n} + \cdots$$
$$= [0; 1, 10, 6, 1, 6, 2, 14, 4, 124, 2, 1, 2, 2039, 1, 9, 1, 1, 1, 262\,111, 2, 8,$$
$$1, 1, 1, 3, 1, 536\,870\,655, 4, 16, 3, 1, 3, 7, 1, 140\,737\,488\,347\,135, \ldots].$$

The increasing sequence of very large partial quotients suggests that something interesting is going on. An explanation was found later by consideration of the continued fractions of an appropriate formal Laurent series and then speciali-sation of the variable to an appropriate integer.

Indeed, van der Poorten and Shallit found experimentally that

$$x^{-1} + x^{-2} + x^{-3} + x^{-5} + \cdots + x^{-F_n} + \cdots$$
$$= [0; x - 1, x^2 + 2x + 2, x^3 - x^2 + 2x - 1, -x^3 + x - 1, -x, -x^4 + x,$$
$$- x^2, -x^7 + x^2, -x - 1, x^2 - x + 1, x^{11} - x^3, -x^3 - x, -x, x,$$
$$x^{18} - x^5, -x, x^3 + 1, x, -x, -x - 1, -x + 1, -x^{29} + x^8, x - 1, \dots].$$

The continued fraction expansion $[0; a_1(x), a_2(x), \dots]$ of a formal power se-ries $f(x) = \sum_{n=1}^{\infty} f_n x^{-n}$, say in $\mathbb{Q}((1/x))$, generically has all its partial quotients $a_n(x)$ of degree 1 (of course, excepting 'early' partial quotients) and coeffi-cients increasing in complexity at a frantic pace – the number of digits of their numerators and denominators seems to increase as $O(n^2)$.

A 'prime' example is

$$x^{-1} + x^{-2} + x^{-3} + x^{-5} + x^{-7} + x^{-11} + x^{-13} + x^{-17} + \cdots$$
$$= [0; x - 1, x^2 + 2x + 2, x - 1, x, x + 1, -x + 3, -\tfrac{1}{6}x - \tfrac{2}{9}, -54x - 36,$$
$$- \tfrac{1}{81}x^7 - \tfrac{2}{81}x, 54x + 36, -\tfrac{1}{30}x + \tfrac{14}{225}, -\tfrac{125}{26}x - \tfrac{5775}{676}, \tfrac{17576}{85625}x - \tfrac{471172}{2346125},$$
$$- \tfrac{1607095625}{21477872}x + \tfrac{111100749375}{1009459984}, \tfrac{5930577406}{1541204704375}x + \tfrac{220062276512}{10788432930625},$$
$$\tfrac{151038061028750}{12125065506567}x - \tfrac{3797528391580000}{117208966563481}, \tfrac{1301767671194406}{528633213600625}x + \tfrac{45843615713376}{10788432930625},$$
$$\tfrac{528633213600625}{8282833201083948}x - \tfrac{4712765099249571875}{23584677303986451601},$$
$$- \tfrac{7761910547758329703}{22731228184826875}x + \tfrac{3501934882619989192}{139634687421079375}, \dots].$$

In contrast,

$$\sum_{n=0}^{\infty} x^{-3^n} = [0; x, -x, -x, -x^3, x, x, -x, -x^9, x, -x, -x,$$
$$x^3, x, x, -x, -x^{27}, x, -x, -x, -x^3, x, x, -x, x^9, x, -x, \dots],$$

which we can deduce from the folding lemma.

Indeed, given a sequence of positive integers $\{G_n\}_{n=0}^{\infty}$ satisfying the inequal-ity $G_{n+1}/G_n > 2$ for all n, the folding lemma entails that a series $\sum_{n=0}^{\infty} x^{-G_n}$ has a folded continued fraction expansion similar to that in the previous sec-tion. Moreover, those function field expansions are so tidy that one can safely and usefully *specialise* them to numerical continued fraction expansions, by replacing the variable x with some integer g of absolute value at least 2. However, the resulting expansions may offend by containing negative-integer partial quotients. All this holds, a little more subtly, for $G_{n+1}/G_n \geq 2$ for all n.

If, however, $G_{n+1}/G_n < 2$ for any, let alone for many, n then it seems that there will be partial quotients with non-integer coefficients and that the non-integrality will propagate in the course of the expansion to become generic behaviour, as in the prime example.

There is, however, an astonishing class of exceptions. Specifically, if $\{G_n\}_n$ is a linear recurrence sequence then we will present compelling evidence that the continued fraction expansion is generic unless, surprisingly, $\{G_n\}_n$ satisfies one of the recurrence relations

$$G_{n+k} = G_{n+k-1} + G_{n+k-2} + \cdots + G_n,$$

with appropriate initial values; these are the Fibonacci numbers, the Lucas numbers and their higher-order immediate generalisations (see Section 1.3). We do not know any other *natural* – thus, not contrived – result that holds for those and only those recurrence sequences.

The secret behind our method is betrayed by the example

$$g(x) = x \sum_{n=0}^{\infty} x^{-2^n} = [1, x, \overline{x}, \overline{x}, \overline{x}, x, x, \overline{x}, \overline{x}, x, \overline{x}, \overline{x}, x, x, x, \ldots]$$

viewed in a different way. Here, obviously,

$$g(x^2) = x(g(x) - 1). \tag{6.1}$$

However (cf. (5.4)), multiplying a continued fraction by x multiplies every second partial quotient by x and divides each remaining partial quotient by x. So, the functional equation (6.1) alleges that

$$[1; x^2, \overline{x}^2, \overline{x}^2, \overline{x}^2, x^2, x^2, \overline{x}^2, \overline{x}^2, x^2, \overline{x}^2, \overline{x}^2, x^2, x^2, x^2, \overline{x}^2, \overline{x}^2, \ldots]$$
$$= [0; 1, \overline{x}^2, \overline{1}, \overline{x}^2, 1, x^2, \overline{1}, \overline{x}^2, 1, \overline{x}^2, \overline{1}, x^2, 1, x^2, \overline{1}, \overline{x}^2, \ldots].$$

Apparently, inserting a ripple $1\overline{1}1\overline{1}1\overline{1}1\overline{1}1\overline{1}1 \cdots$ or, if one prefers, a *fold* of 1s into a continued fraction expansion, and also *rippling* – changing the sign of alternate – pre-existing entries, needs no more than a change in the zeroth entry from 1 to 0. This, indeed, is the ripple lemma re-alleged below.

6.4 Folding and rippling

Suppose that we know the continued fraction expansion for $S(n)$, where

$$S(n) = x^{G_m}(x^{-G_m} + x^{-G_{m+1}} + x^{-G_{m+2}} + \cdots + x^{-G_{m+n}}).$$

We step our knowledge up to $S(n + 1)$ by the following strategy:

1. replace m by $m + 1$;
2. divide by $x^{G_{m+1}-G_m}$;
3. add 1.

Of course, we do know that

$$S(0) = 1 = [1], \qquad S(1) = [1, x^{G_{m+1}-G_m}];$$

therefore this process has a beginning. After sufficiently many steps, we divide by x^{G_m} and will then have obtained the continued fraction expansion of

$$S = x^{-G_m} + x^{-G_{m+1}} + x^{-G_{m+2}} + x^{-G_{m+3}} + x^{-G_{m+4}} + \cdots.$$

Recall the rule (5.4) in the form

$$M[Na, Mb, Nc, Md, Ne, \ldots] = N[Ma, Nb, Mc, Nd, Me, \ldots],$$

incisively illustrating how multiplying, or dividing, divides or respectively multiplies every second partial quotient. Unfortunately, however, our expansions begin with the partial quotient 1, which is not divisible by much at all. If only that 1 were 0 – because zero is divisible by anything! Happily, we have the following:

Lemma 6.8 (Ripple lemma [128]) *For space-saving reasons write $\overline{x}$ for $-x$. Then*

$$[1, a, b, c, d, e, \ldots] = [0, 1, \overline{a}, \overline{1}, b, 1, \overline{c}, \overline{1}, d, 1, \overline{e}, \overline{1}, \ldots].$$

Specifically, terminating versions of the lemma are given by

$$[1, a, \beta] = [0, 1, \overline{a} - 1, \beta] \qquad \text{and} \qquad [1, a, b, \gamma] = [0, 1, \overline{a}, \overline{1}, b + 1, \gamma].$$

Amusingly, the ripple lemma is an immediate consequence of a simple rule for changing the sign of partial quotients in a continued fraction expansion. As example, set $G_n = 2^n$, noting that $G_{m+1} - G_m = G_m$. Recall we know that $S(1) = x^{G_m}(x^{-G_m} + x^{-G_{m+1}}) = [1, x^{G_m}]$.

We will loop, or iterate, as follows:

(a) ripple the continued fraction expansion and replace m by $m + 1$;
(b) divide by $x^{G_{m+1}-G_m} = x^{G_m}$ and add 1.

We get (a) $[0, 1, \overline{x}^{G_{m+1}}, \overline{1}]$ and (b) $[1, x^{G_m}, \overline{x}^{G_m}, \overline{x}^{G_m}]$. The next loop gives (a) $[0, 1, \overline{x}^{G_{m+1}}, \overline{1}, \overline{x}^{G_{m+1}}, 1, x^{G_{m+1}}, \overline{1}]$ and (b) $[1, x^{G_m}, \overline{x}^{G_m}, \overline{x}^{G_m}, \overline{x}^{G_m}, x^{G_m}, x^{G_m}, \overline{x}^{G_m}]$. Plainly, each loop doubles the number of stable entries of the expansion: we started with one, then had two and now have four stable entries. Yet more clearly, every entry after the zeroth will be $\pm x^{G_m}$. A little less obviously, but evident if one has the correct experience, the sequence of signs $\pm$ is a *paper-folding sequence* given by the pattern of creases in a piece of paper folded repeatedly in half.

We obtain the expansion of $x^{G_m} \sum_{n \geq m} x^{-G_n}$ and, finally, divide by x^{G_m}.

For general G_m, it is easy to check that after one loop we have an entry $x^{G_{m+1}-G_m}$, after a second loop the exponent $G_{m+2} - 2G_{m+1} + G_m$ also appears, after three loops we also see the exponent $G_{m+3} - 2G_{m+2} + G_m$ etc. After n loops the oldest exponent is $G_{m+n} - 2G_{m+n-1} + G_m$.

Case A. In brief, the argument just now sketched works flawlessly if the oldest exponent $G_{m+n} - 2G_{m+n-1} + G_m$ is positive for all n; we will refer to this desirable situation as case A. At the very least a 'plan B' will be required if $G_{m+k+1} - 2G_{m+k} + G_m = 0$ for some k (case B_k below), and it is very likely to be a bad, bad thing if after $n - 1$ loops the oldest exponent is positive but that exponent becomes negative after one more loop.

The good news is that in fact there is an obvious plan B that applies if $G_{m+k+1} - 2G_{m+k} + G_m = 0$. This relation defines the Fibonacci and Lucas numbers if $k = 2$, and their generalisations for greater k. By the way, though more general results could be obtained from the arguments above, here and from now on we suppose that $\{G_n\}_n$ is a *linear recurrence sequence*.

In the special example $G_n = 2^n$ the final division leads to the expansion

$$[0, x^{G_m}, \overline{1}, \overline{x}^{G_m}, \overline{1}, x^{G_m}, 1, \overline{x}^{G_m}, \overline{1}, x^{G_m}, \overline{1}, \overline{x}^{G_m}, 1, x^{G_m}, 1, \ldots],$$

but entries in such an expansion should be polynomials of degree at least 1.

Analogously, the expansion $-\pi = [\overline{3}, \overline{7}, \overline{15}, \overline{1}, \overline{292}, \overline{1}, \ldots]$ is filled with *in-admissible* negative entries. But in fact $-\pi = [\overline{4}, 1, 6, 15, 1, 292, 1, \ldots]$ and so one should be able to obtain that directly from the inadmissible result. Indeed, to do this one notices that, above, negating a negation produced a spontaneous partial quotient 1; hence the ripple lemma.

From Example 4.21 we have

$$-\beta = [0; 1, \overline{1}, 1, 0, \beta],$$

and from Exercise 4.22 we have

$$-\beta = [0; \overline{1}, 1, 1, \overline{1}, 0, \beta].$$

Further, recall from (4.3) that

$$[\ldots, a, 0, b, \ldots] = [\ldots, a + b, \ldots].$$

Then

$$-\pi = [\overline{3}; 0, \overline{1}, 1, \overline{1}, 0, 7, 15, 1, 292, \ldots] = [\overline{4}; 1, 6, 15, 1, 292, \ldots].$$

We see that the final division in the example $G_n = 2^n$, $m = 0$, leads to

$$[0; x, \bar{1}, \bar{x}, \bar{1}, x, 1, \bar{x}, \bar{1}, x, \bar{1}, \bar{x}, 1, x, 1, \ldots].$$

Applying our removal of ± 1s by the negation strategy gives

$$[0; x, \bar{1}, 0, 1, \bar{1}, 1, 0, x, 1, \bar{x}, \bar{1}, x, 1, \bar{x}, 1, x, \bar{1}, \bar{x}, \bar{1}, \ldots]$$

$$= [0; x-1, x+1, 1, 0, \bar{1}, 1, \bar{1}, 0, x, 1, \bar{x}, \bar{1}, x, \bar{1}, \bar{x}, 1, x, 1, \ldots]$$

$$= [0; x-1, x+2, x-1, 1, 0, \bar{1}, 1, \bar{1}, 0, x, 1, \bar{x}, 1, x, \bar{1}, \bar{x}, \bar{1}, \ldots]$$

$$= [0; x-1, x+2, x, x-1, 1, 0, \bar{1}, 1, \bar{1}, 0, x, \bar{1}, \bar{x}, 1, x, 1, \ldots]$$

$$\vdots$$

$$= [0; x-1, x+2, x, x, x-2, x, x+2, x, \ldots] = [0; x-1, S^{\infty}_{x,x-2}(x+2, x)].$$

Here $S_p(w) = wp^tw$ is the *perturbed symmetry operator* on a word w. All this is trivial algorithmically, but it is painful to do by hand (and to type!).

Case B_k: $G_{m+k+1} - 2G_{m+k} + G_m = 0$. For temporary convenience we write $G_{(n)}$ in place of $G_{m+n} - 2G_{m+n-1} + G_m$ to emphasise that a partial quotient $x^{G_{(n)}}$ is n loops old. For example, after three loops we have

$$S(3) = x^{G_m}(x^{-G_m} + x^{-G_{m+1}} + x^{-G_{m+2}} + x^{-G_{m+3}})$$

$$= [1, x^{G_{(1)}}, \bar{x}^{G_{(2)}}, \bar{x}^{G_{(1)}}, \bar{x}^{G_{(3)}}, x^{G_{(1)}}, x^{G_{(2)}}, \bar{x}^{G_{(1)}}],$$

and if we begin the next loop with a partial ripple then we also have

$$S(3) = [0, 1, \bar{x}^{G_{(1)}}, \bar{1}, \bar{x}^{G_{(2)}}, 1, x^{G_{(1)}}, \bar{1}, \bar{x}^{G_{(3)}} + 1, x^{G_{(1)}}, x^{G_{(2)}}, \bar{x}^{G_{(1)}}].$$

Indeed, in the special case $G_n = F_n$, we have for the Fibonacci numbers $F_{n+2} = F_{n+1} + F_n$. Thus case B_2 applies, we have $G_{(3)} = 0$ and we may forthwith complete the loop to obtain

$$S(4) = [1, x^{G_{(1)}}, \bar{x}^{G_{(2)}}, \bar{x}^{G_{(1)}}, \bar{x}^{G_{(3)}}, x^{G_{(1)}}, x^{G_{(2)}}, \bar{x}^{G_{(1)}}, 0, x^{F_{m+1}}, x^{G_{(3)}}, \bar{x}^{F_{m+1}}].$$

Finally, 'eating the zero' yields

$$S(4) = [1, x^{G_{(1)}}, \bar{x}^{G_{(2)}}, \bar{x}^{G_{(1)}}, \bar{x}^{G_{(3)}}, x^{G_{(1)}}, x^{G_{(2)}}, \bar{x}^{G_{(1)}} + x^{F_{m+1}}, x^{G_{(3)}}, \bar{x}^{F_{m+1}}].$$

The next loop requires that we first insert two partial ripples, giving

$$S(4) = [0, 1, \bar{x}^{G_{(1)}}, \bar{1}, \bar{x}^{G_{(2)}}, 1, x^{G_{(1)}}, \bar{1}, \bar{x}^{G_{(3)}} + 1,$$
$$x^{G_{(1)}}, x^{G_{(2)}}, \bar{x}^{G_{(1)}} + x^{F_{m+1}}, 0, x^{G_{(3)}}, x^{F_{m+1}}, \bar{1}],$$

and we then complete the loop to obtain

$$S(5) = [1, x^{G_{(1)}}, \bar{x}^{G_{(2)}}, \bar{x}^{G_{(1)}}, \bar{x}^{G_{(3)}}, x^{G_{(1)}}, x^{G_{(2)}}, \bar{x}^{G_{(1)}}, 0, x^{F_{m+1}}, x^{G_{(3)}},$$
$$\bar{x}^{F_{m+1}} + x^{F_{m+2}+F_{m-1}}, 0, x^{G_{(1)}}, x^{2F_m}, \bar{x}^{G_{(1)}}].$$

It is not yet totally obvious that the expansion has stabilised and is growing steadily in length. However, the parity of the occurrences of $\pm x^{G_{(3)}} = \pm 1$ remains friendly to our cause; the same will hold for $x^{G_{(k+1)}}$ in the case B_k.

Back to work, we first have

$$S(5) = [0, 1, \overline{x}^{G_{(1)}}, \overline{1}, \overline{x}^{G_{(2)}}, 1, x^{G_{(1)}}, \overline{1}, \overline{x}^{G_{(3)}} + 1, x^{G_{(1)}}, x^{G_{(2)}},$$
$$\overline{x}^{G_{(1)}} + x^{F_{m+1}}, 0, x^{G_{(3)}}, x^{F_{m+1}} + \overline{x}^{F_{m+2}+F_{m-1}} + \overline{x}^{G_{(1)}}, \overline{1}, x^{2F_m}, 1,$$
$$x^{G_{(1)}}, \overline{1}]$$

and, completing the loop,

$$S(6) = [1, x^{G_{(1)}}, \overline{x}^{G_{(2)}}, \overline{x}^{G_{(1)}}, \overline{x}^{G_{(3)}}, x^{G_{(1)}}, x^{G_{(2)}}, \overline{x}^{G_{(1)}} + x^{F_{m+1}}, x^{G_{(3)}},$$
$$\overline{x}^{F_{m+1}} + x^{F_{m+2}+F_{m-1}} + x^{G_{(1)}}, x^{2F_m} + \overline{x}^{F_{m+2}+2F_m} + \overline{x}^{G_{(2)}},$$
$$\overline{x}^{G_{(1)}}, x^{F_{m+2}}, x^{G_{(1)}}, x^{G_{(2)}}, \overline{x}^{G_{(1)}}].$$

We remark that in case B_k the exponents do not age smoothly as they do in case A. First, their expected behaviour is that they decrease (rather than non-strictly increase). Second, the vanishing of $G_{(k+1)}$ may change the positional parity of exponents, leading them to grow exponentially rather than to politely shrink away. Notice that the expansion is struggling to display some folded symmetry and surely will succeed in doing so after the next loop. Several initial partial quotients are now fixed.

Indeed, it is now safe to remark that

$$S(6) = [\ldots, \overline{x}^{G_{(1)}} + x^{F_{m+1}}, 0, x^{G_{(3)}}, x^{F_{m+1}} + \overline{x}^{F_{m+2}+F_{m-1}} + \overline{x}^{G_{(1)}}, \overline{1},$$
$$x^{2F_m} + \overline{x}^{F_{m+2}+2F_m} + \overline{x}^{G_{(2)}}, 1, x^{G_{(1)}}, \overline{1}, x^{F_{m+2}}, 1, \overline{x}^{G_{(1)}}, \overline{1}, x^{G_{(2)}}, 1, x^{G_{(1)}}, \overline{1}]$$

and, therefore, filling in the ellipses,

$$S(7) = [1, x^{G_{(1)}}, \overline{x}^{G_{(2)}}, \overline{x}^{G_{(1)}}, \overline{x}^{G_{(3)}}, x^{G_{(1)}}, x^{G_{(2)}}, \overline{x}^{F_{m+1}} + x^{F_{m+2}+F_{m-1}} + x^{G_{(1)}},$$
$$x^{2F_m} + \overline{x}^{F_{m+2}+2F_m} + \overline{x}^{G_{(2)}}, \overline{x}^{G_{(1)}}, x^{F_{m+2}} + \overline{x}^{F_{m+4}} + \overline{x}^{G_{(3)}}, x^{G_{(1)}}, x^{G_{(2)}},$$
$$\overline{x}^{G_{(1)}}, x^{F_{m+2}+F_m}, x^{G_{(1)}}, \overline{x}^{G_{(2)}}, \overline{x}^{G_{(1)}}, x^{G_{(3)}}, x^{G_{(1)}}, x^{G_{(2)}}, \overline{x}^{G_{(1)}}],$$

nicely displaying a (highly perturbed) folding.

We can now announce what is in effect the main lemma of this chapter: surprisingly, the 22 partial quotients comprising $S(7)$ are precisely the partial quotients commencing the expansion of $S(\infty)$.

It is now time to recall that $G_{(1)} = F_{m-1}$, $G_{(2)} = F_{m-2}$, $G_{(3)} = 0$, thus obtaining, when $k = 2$,

$$S(\infty) = [1, x^{F_{m-1}}, \overline{x}^{F_{m-2}}, \overline{x}^{F_{m-1}}, \overline{x}^0, x^{F_{m-1}}, x^{F_{m-2}},$$

$$\overline{x}^{F_{m+1}} + x^{F_{m+2}+F_{m-1}} + x^{F_{m-1}}, x^{2F_m} + \overline{x}^{F_{m+2}+2F_m} + \overline{x}^{F_{m-2}},$$

$$\overline{x}^{F_{m-1}}, x^{F_{m+2}} + \overline{x}^{F_{m+4}} + \overline{x}^0, x^{F_{m-1}}, x^{F_{m-2}}, \overline{x}^{F_{m-1}},$$

$$x^{F_{m+2}+F_m}, x^{F_{m-1}}, \overset{\leftarrow}{\overline{x}^{F_{m-2}}}, \overline{x}^{F_{m-1}}, x^0, x^{F_{m-1}}, x^{F_{m-2}}, \overline{x}^{F_{m-1}}, \dots],$$

where the arrow denotes a reversal.

We have retained the general notation in an effort to emphasise that our remarks apply not just to the Fibonacci and Lucas numbers but generally to their higher order generalisations. The point is that the particular case $k = 2$ amply illustrates the complexity of the more general one and that for each k there is a minimal $n = n(k)$ so that the continued fraction expansion of $S(n)$ is a prefix of that of $S(\infty)$. Here $n(2) = 7$.

Case C: A bad, bad, thing. Experimentally, one sees that almost any mistype or omission in giving data to a symbolic algebra system (such as PARI-GP) seems invariably to yield a generic continued fraction expansion: the logarithmic height of the coefficients of the partial quotients grows at a quadratic pace. Why that's so is explained below.

Recall that rippling is made possible by the identities

$$[A + 1, B, \gamma] = [A + 1, 0, \overline{1}, 1, \overline{1}, 0, \overline{B}, \overline{\gamma}] = [A, 1, \overline{B + 1}, \overline{\gamma}].$$

Therefore, for an arbitrary nonzero constant c, we have

$$[A + c, B, \gamma] = c[A/c + 1, cB, \gamma/c] = c[A/c, 1, \overline{cB + 1}, \overline{\gamma}/c]$$

$$= [A, 1/c, \overline{c^2 B + c}, \overline{\gamma}/c^2].$$

If $c = \pm 1$ this makes a mess which propagates through the continued fraction expansion precisely in such a way as to produce the generic quadratic growth in the size of the numerical coefficients of the partial quotients.

If $G_{m+n+1} - 2G_{m+n} + G_m$ is negative and $G_{m+n} - 2G_{m+n-1} + G_m$ is positive then the relevant loop will first require division by some power of x smaller than the $(G_{m+1} - G_m)$th, followed by another ripple and division by the remaining power of x. That 'further ripple' will seemingly involve moving some constant c different from ± 1, very likely provoking a propagating 'mess' as suggested above, although we have no firm proof.

Van der Poorten was interested to discover that actually the $S(n)$ are fairly tolerant of error. He experimented on

$$S(n) = x^5(x^{-5} + x^{-8} + x^{-13} + x^{-21} + \cdots + x^{-F_{n+5}})$$

and thought he might strike lucky if he replaced the exponent $F_7 = 13$ by 14. That made a smaller mess than van der Poorten had expected (its propagation was rather muted), but dividing that by x^5 provided 47 pages of chaos. However, while $S(n)$ always has a good continued fraction expansion, the series

$$x^{-3} + x^{-5} + x^{-8} + x^{-13} + x^{-21} + \cdots,$$

for example, does *not* have a specialisable continued fraction expansion. If $G_m \neq 1$, one cannot be sure of being able to safely divide the expansion for $S(n)$ by x^{G_m}, further restricting the class of specialisable examples.

Notes

A detailed discussion of the continued fraction expansion of binary decimals $s_a = 2\sum_{n=0}^{\infty}(-1)^{a_n}2^{-2^n}$, $a = 0.a_1a_2a_3\ldots$, is to be found in [134].

Our proof of Zaremba's conjecture for powers of 2 is a simplified version of a result in [117].

Computations by Shallit provoked the conjecture discussed in this chapter; a proof of the cases B_2 and B_3 by a rather more intricate method than that suggested here can be found in [135]. Finally, in [128] van der Poorten discussed the subject generally, explained the ripple lemma and proposed the argument detailed above for the cases B_k.

For more on specialisability, see [111].

7

The integer part of $q\alpha + \beta$

We turn to the study of a class of generating functions that, somewhat like the folds and ripples of the previous chapter, lead to remarkable continued fractions and rational approximations. They rely on an inhomogeneous continuous function algorithm discussed below.

7.1 Inhomogeneous Diophantine approximation

As we have already learnt, the convergents p_n/q_n of a continued fraction $[a_0; a_1, a_2, \ldots]$ representing the real *irrational* number α minimise the quantity $|q\alpha - p|$. The related *homogeneous* Diophantine problem initiated by Dirichlet's theorem (Theorem 1.36) and discussed in Sections 1.4 and 1.5 was in fact our motivation to develop the theory of continued fractions.

It is not therefore completely unreasonable to believe that a slightly more general minimisation problem for the quantity $|q\alpha + \beta - p|$, where β is another (not necessarily irrational) real number already considered in Chebyshev's theorem (Exercise 1.39), gives rise to a natural extension of continued fractions. More important is not so much the algorithmic solution of this inhomogeneous Diophantine approximation problem but its many consequences.

Because the replacement of α and β by their fractional parts does not affect the Diophantine problem, we will assume that they lie between 0 and 1.

EXTENDED CONTINUED FRACTION ALGORITHM Suppose that $\alpha, \beta \in [0, 1)$ and α is irrational.

Let $\alpha_0 = \alpha, \beta_0 = \beta$, and inductively define, for $n = 0, 1, 2, \ldots$,

$$a_{n+1} = \left\lfloor \frac{1}{\alpha_n} \right\rfloor, \qquad b_{n+1} = \left\lfloor \frac{\beta_n}{\alpha_n} \right\rfloor,$$

$$\alpha_{n+1} = \frac{1}{\alpha_n} - a_{n+1}, \qquad \beta_{n+1} = \frac{\beta_n}{\alpha_n} - b_{n+1}.$$

Finally, introduce the 'generalised' convergents

$$p_{n+1} = a_{n+1}p_n + p_{n-1}, \quad p_0 = 0, \ p_{-1} = 1,$$

$$q_{n+1} = a_{n+1}q_n + q_{n-1}, \quad q_0 = 1, \ q_{-1} = 0,$$

$$r_{n+1} = a_{n+1}r_n + r_{n-1} + b_{n+1}, \quad r_0 = 0, \ r_{-1} = 0,$$

$$s_n = (a_{n+1} + (-1)^{n+1}b_{n+1})p_n + s_{n-1}, \quad s_{-1} = 1,$$

$$t_n = (a_{n+1} + (-1)^{n+1}b_{n+1})q_n + t_{n-1}, \quad t_{-1} = 1$$

and define the integers $\widehat{p}_n$ and $\widehat{q}_n$ by

$$(\widehat{p}_{2k}, \widehat{q}_{2k}) = \begin{cases} (s_{2k}, t_{2k}) & \text{if } b_{2k+2} = 0 \\ & \text{and } t_{2k} \le q_{2k} + q_{2k+1}, \\ (s_{2k} - p_{2k}, t_{2k} - q_{2k}) & \text{otherwise;} \end{cases}$$

$$(\widehat{p}_{2k+1}, \widehat{q}_{2k+1}) = \begin{cases} (s_{2k+1}, t_{2k+1}) & \text{if } b_{2k+2} = 0 \\ & \text{and } t_{2k+1} \le q_{2k+1} + q_{2k+2}, \\ (s_{2k+1} - p_{2k+2}, t_{2k+1} - q_{2k+2}) & \text{otherwise.} \end{cases}$$

This concludes the algorithm.

Note that $\{a_n\}_n$, $\{p_n\}_n$, and $\{q_n\}_n$ are the familiar quantities that arise in the continued fraction expansion of

$$\alpha = [a_1, a_2, \dots] = [0; a_1, a_2, \dots].$$

However, the sequence $\{\alpha_n\}_{n \ge 0}$ so constructed represents the *reciprocals* of the complete quotients from Chapter 2 instead of the quotients themselves. This change is made for the following reason: the quantities

$$\alpha_n = [a_{n+1}, a_{n+2}, \dots] = [0; a_{n+1}, a_{n+2}, \dots],$$

along with the quantities β_n, all lie in the unit interval $[0, 1)$.

The next lemmata examine some properties of the algorithm. In particular, they imply that the sequences $\{\widehat{p}_n\}_n$ and $\{\widehat{q}_n\}_n$ solve the Chebyshev–Minkowski problem (a sharper version of Exercise 1.39), which is to find p, q such that

$$|q\alpha + \beta - p| < \frac{1}{q}. \tag{7.1}$$

Lemma 7.1 (cf. Lemma 2.11) *For all $n \geq 0$,*

$$\alpha = \frac{\alpha_{n+1}p_n + p_{n+1}}{\alpha_{n+1}q_n + q_{n+1}} \quad and \quad \beta = \frac{\alpha_{n+1}r_n + r_{n+1} + \beta_{n+1}}{\alpha_{n+1}q_n + q_{n+1}}.$$

Proof We see that the first relation is equivalent to the one from Example 2.36 by noting that here the complete quotients are just the reciprocals of those from Chapter 2. The second relation is a companion to the first and follows from the relations

$$\frac{1}{\alpha_n} = \alpha_{n+1} + a_{n+1}, \qquad \frac{\beta_n}{\alpha_n} = \beta_{n+1} + b_{n+1}$$

and an easy inductive argument. □

Now let $\delta_n = q_n\alpha - p_n$ for $n = 0, 1, \ldots$

Lemma 7.2 (cf. Lemma 2.24) *For all $n \geq 0$, we have*

$$\delta_n = \frac{(-1)^n}{\alpha_{n+1}q_n + q_{n+1}}, \qquad \delta_{n+1} = \frac{(-1)^{n+1}\alpha_{n+1}}{\alpha_{n+1}q_n + q_{n+1}}$$

and

$$t_n\alpha + \beta - s_n = \frac{(-1)^n(1 - \alpha_{n+1}) + \beta_{n+1}}{\alpha_{n+1}q_{n+1} + q_n}$$

$$= \delta_n + \delta_{n+1} + (-1)^n\delta_n\beta_{n+1}.$$

Proof It is sufficient to establish the last relation. By Lemma 7.1 we have

$$t_n\alpha + \beta - s_n = \frac{u_n + \alpha_{n+1}v_n + \beta_{n+1}}{\alpha_{n+1}q_n + q_{n+1}},$$

where $u_n = t_np_{n+1} - s_nq_{n+1} + r_{n+1}$ and $v_n = t_np_n - s_nq_n + r_n$. Now we check that $u_{-1} = -1$, $u_0 = 1$, $v_{-1} = 1$, $v_0 = -1$ and that

$$v_{n+1} = u_n \quad and \quad u_n = a_{n+1}(v_n + (-1)^n) + v_{n-1}.$$

It now follows inductively that $u_n = (-1)^n$ and $v_n = (-1)^{n+1}$. □

Lemma 7.3 *There exists m such that $a_{2m+1} > b_{2m+1}$. More precisely, if $b_{2k+1} = a_{2k+1}$ for $k = 0, 1, \ldots, m$ then $q_{2m+3} < 2/(1-\beta)$.*

Proof If $b_{2k+1} = a_{2k+1}$ for $k \leq m$ then $b_{2k+2} = 0$ for those k; hence

$$0 = t_{-1} - q_0 = t_0 - q_0 = \cdots = t_{2k-1} - q_{2k}$$

$$= t_{2k} - q_{2k} - q_{2k}(a_{2k+1} - b_{2k+1}) = t_{2k} - q_{2k}$$

$$= t_{2k+1} - q_{2k+2} - b_{2k+2}q_{2k+1} = t_{2k+1} - q_{2k+2}.$$

Similarly,

$$0 = s_{-1} - p_0 = s_0 - p_0 = \cdots = s_{2k} - p_{2k} = s_{2k+1} - p_{2k+2};$$

thus $t_{2k+1} = q_{2k+2}$ and $s_{2k+1} = p_{2k+2} + 1$. It then follows from Lemma 7.2 that

$$1 - \beta = 2\delta_{2k+2} + (1 - \beta_{2k+1})\delta_{2k+1} \leq 2\delta_{2k+2}$$

for all $k \leq m$, so that $1 - \beta < 2/q_{2m+3}$. $\qquad\qquad\qquad\qquad\qquad$ □

Since $q_{2m+3} \geq F_{2m+4}$ for the Fibonacci numbers (see Section 1.3) and $F_4 = 3$, $F_6 = 8$ we deduce that $\beta \leq 3/4$ implies that either $a_1 > b_1$ or $a_3 > b_3$.

For any real but fixed $\delta \neq 0$, there are infinitely many solutions in positive integers N of the inequality

$$\lfloor N\alpha + \beta + \delta \rfloor \neq \lfloor N\alpha + \beta \rfloor;$$

this follows from the uniform distribution of the fractional parts of $N\alpha$ for α irrational [90]. In particular there is always a minimal such solution, which we will call $N(\delta)$. Furthermore, we take $N_n = N(\delta_n)$ where, as before, $\delta_n = q_n\alpha - p_n$.

Lemma 7.4 *Given a nonnegative integer n, let $N' > N'' \geq 1$ be two solutions of the inequality*

$$\lfloor N\alpha + \beta + \delta_n \rfloor \neq \lfloor N\alpha + \beta \rfloor. \qquad\qquad (7.2)$$

Then $N' - N'' = kq_{n+1} + lq_n$ for some integers $k \geq 1$ and $l \geq 0$.

Proof We will establish only the case of even n (when $\delta_n > 0$); the odd-n case is similar. The hypothesis implies the existence of two integers M' and M'' such that

$$N'\alpha + \beta + \delta_n \geq M' > N'\alpha + \beta \qquad \text{and} \qquad N''\alpha + \beta + \delta_n \geq M'' > N''\alpha + \beta.$$

Setting $N = N' - N'' > 0$ and $M = M' - M''$ and using $\alpha > p_n/q_n$ (as n is even), we find

$$(N + q_n)\delta_n > Mq_n - Np_n > -\delta_n q_n.$$

Take $k = Mq_n - Np_n$, so that $k \geq 0$ from the estimate $|\delta_n| < 1/q_{n+1}$. If $k = 0$ then $M' - M'' = M = lp_n$ and $N' - N'' = N = lq_n$ for some integer $l \geq 1$, so that $M'' + lp_n > (N'' + lq_n)\alpha + \beta$; hence

$$M'' > N''\alpha + \beta + l\delta_n \geq N''\alpha + \beta + \delta_n \geq M'',$$

a contradiction. Thus $k \geq 1$. It follows now from $p_{n+1}q_n - q_{n+1}p_n = 1$ that

$$Mq_n - Np_n = k = (kp_{n+1})q_n - (kq_{n+1})p_n.$$

As $n > 0$ and p_n, q_n are relatively prime, the preceding equation implies that $M = kp_{n+1} + lp_n$ and $N = kq_{n+1} + lq_n$ for some integer l. But $(N + q_n)\delta_n > k$ yields $N > kq_{n+1} - q_n$, so that l must be nonnegative (even for $n = 0$). $\qquad$ □

Lemma 7.5 *If $N_{n+1} \leq q_n$ then $N_n = N_{n+1} + q_{n+1}$. In particular, either $N_n + q_n$ or $N_{n+1} + q_{n+1}$ exceeds $q_n + q_{n+1}$.*

Proof As in the previous proof, we concentrate only on the case of even n. Since N_{n+1} is (the least) positive solution of $\lfloor N\alpha + \beta + \delta_{n+1} \rfloor \neq \lfloor N\alpha + \beta \rfloor$, there exists M such that

$$N_{n+1}\alpha + \beta + \delta_{n+1} < M \leq N_{n+1}\alpha + \beta.$$

Using $\delta_n > -\delta_{n+1}$ we can derive

$$(N_{n+1} + q_{n+1})\alpha + \beta < M + p_{n+1} < (N_{n+1} + q_{n+1})\alpha + \beta + \delta_n.$$

Hence $N = N_{n+1} + q_{n+1}$ satisfies (7.2) and so either $N_{n+1} + q_{n+1} = N_n$, as claimed, or $N_{n+1} + q_{n+1} - N_n = kq_{n+1} + lq_n$ for some integers $k \geq 1$ and $l \geq 0$ by Lemma 7.4. Now the conditions $N_{n+1} \leq q_n$ and $N_n \geq 1$ imply that $l = 0$ and $k = 1$ or, equivalently, that $N_{n+1} = N_n$. But then there exist integers M' and M'' such that

$$N_n\alpha + \beta + \delta_{n+1} = N_{n+1}\alpha + \beta + \delta_{n+1} < M' \leq N_{n+1}\alpha + \beta$$
$$= N_n\alpha + \beta < M'' \leq N_n\alpha + \beta + \delta_n,$$

so that $\delta_n - \delta_{n+1} > M'' - M' \geq 1$, which is impossible. □

Corollary 7.6 (Best one-sided approximation) *If $q\alpha + \beta$ is never an integer, the inequalities*

$$\{q\alpha + \beta\} < \{(N_n + q_n)\alpha + \beta\} \qquad \textit{for } n \textit{ even,}$$
$$\{q\alpha + \beta\} > \{(N_n + q_n)\alpha + \beta\} \qquad \textit{for } n \textit{ odd}$$

are impossible for the range $q_n < q < q_{n-1} + N_n$.

In particular, $(N_n + q_n)\alpha + \beta$ is the best one-sided approximation $q\alpha + \beta$ to an integer in the range $q_n < q < q_n + q_{n+1} + N_n$.

Proof As in the proof of Lemma 7.5 we first check that $N = q - q_n$ satisfies $\lfloor N\alpha + \beta \rfloor \neq \lfloor N\alpha + \beta + \delta_n \rfloor$ and then apply Lemma 7.4. Hence either $q = q_n + N_n$ or $q = kq_{n+1} + (l+1)q_n + N_n$ for integers $k \geq 1$, $l \geq 0$, and the conclusion follows. □

In particular, if $N_n \leq q_{n+1}$ then $N_n + q_n$ gives the best one-sided approximation to an integer for $q_n < q \leq q_n + q_{n+1}$.

For the remaining statements in this section we choose m to be the smallest nonnegative integer for which $a_{2m+1} > b_{2m+1}$; its existence was established in Lemma 7.3.

Lemma 7.7 *For each $k \geq 0$, we have $t_{2k} \geq q_{2k}$ and $t_{2k+1} \geq q_{2k+2}$, and these inequalities are strict when $k \geq m$.*

Proof As in the proof of Lemma 7.3, $t_{2k} = q_{2k}$ and $t_{2k+1} = q_{2k+2}$ for $k < m$. Then, for $k \geq m$,

$$t_{2k} = t_{2k-1} + (a_{2k+1} - b_{2k+1})q_{2k} > q_{2k}$$

and

$$t_{2k+1} - q_{2k+2} \geq t_{2k} - q_{2k} > 0. \qquad \qquad \square$$

Lemma 7.8 *For all n and the numbers $\widehat{q}_n$ defined in the extended continued fraction algorithm, one has $\widehat{q}_n \geq q_n$. Moreover, the latter inequality is strict for $n \geq 2m + 1$.*

Proof We will establish this inductively, treating the pairs $n = 2k$ and $n = 2k + 1$ separately.

First consider the case $b_{2k+2} > 0$. This inequality implies that $a_{2k+1} > b_{2k+1}$ (if $a_{2k+1} = b_{2k+1}$ then $\alpha_{2k+1} - \beta_{2k+1} = (1 - \beta_{2k})/\alpha_{2k} > 0$, so that $\beta_{2k+1}/\alpha_{2k+1} < 1$ and $b_{2k+2} = \lfloor \beta_{2k+1}/\alpha_{2k+1} \rfloor = 0$, a contradiction). Therefore, $k \geq m$ and

$$\widehat{q}_{2k} = t_{2k} - q_{2k} = t_{2k-1} - q_{2k} + (a_{2k+1} - b_{2k+1})q_{2k}$$
$$\geq t_{2k-1} - q_{2k} + q_{2k} \geq q_{2k},$$

where the latter inequality is strict for $k \geq m + 1$ by Lemma 7.7. Also, if $b_{2k+2} > 0$ then

$$\widehat{q}_{2k+1} = \widehat{q}_{2k} + c_{2k+2}q_{2k+1} > q_{2k+1}.$$

Our second case is $b_{2k+2} = 0$ and $t_{2k} > q_{2k} + q_{2k+1}$. The latter condition is equivalent to $t_{2k+1} > q_{2k+1} + q_{2k+2}$. Therefore,

$$\widehat{q}_{2k} = t_{2k} - q_{2k} > q_{2k} \qquad \text{and} \qquad \widehat{q}_{2k+1} = t_{2k+1} - q_{2k+2} > q_{2k+1}.$$

Finally, consider the case $b_{2k+2} = 0$ and $t_{2k} \leq q_{2k} + q_{2k+1}$. Then

$$\widehat{q}_{2k} = t_{2k} = t_{2k-1} + (a_{2k+1} - b_{2k+1})q_{2k} \geq q_{2k},$$

with strict inequality for $k > m$, and

$$\widehat{q}_{2k+1} = t_{2k+1} = t_{2k} + (a_{2k+2} + c_{2k+2})q_{2k+1} = t_{2k} - q_{2k} + q_{2k+2}$$
$$= \widehat{q}_{2k} - q_{2k} + q_{2k+2} > q_{2k+2}. \qquad \qquad \square$$

Lemma 7.9 *For $n \geq -1$ we have $\widehat{q}_n \leq q_n + q_{n+1}$.*

Proof Again we distinguish the parity of n and proceed by cases inductively. Note that $t_{-1} = 1 = q_{-1}$; hence $\widehat{q}_{-1} \leq q_{-1} + q_0$. Below we show that $\widehat{q}_{2k-1} \leq q_{2k-1} + q_{2k}$ implies the result for $n = 2k$ and $2k + 1$.

If $b_{2k+2} > 0$ then

$$\widehat{q}_{2k} = t_{2k} - q_{2k} = t_{2k-1} - q_{2k} + (a_{2k+1} - b_{2k+1})q_{2k}$$
$$\leq t_{2k-1} - q_{2k} + q_{2k+1} - q_{2k-1} = \widehat{q}_{2k-1} - q_{2k-1} + q_{2k+1}$$
$$\leq q_{2k} + q_{2k+1},$$

and similarly

$$\widehat{q}_{2k+1} = t_{2k+1} - q_{2k+2} = t_{2k} - q_{2k} + b_{2k+2}q_{2k+1}$$
$$\leq \widehat{q}_{2k} + a_{2k+2}q_{2k+1} \leq (q_{2k} + q_{2k+1}) + q_{2k+2} - q_{2k}$$
$$= q_{2k+1} + q_{2k+2}.$$

If $b_{2k+2} = 0$ and $t_{2k} \leq q_{2k} + q_{2k+1}$ then $t_{2k+1} \leq q_{2k+1} + q_{2k+2}$, so that $\widehat{q}_{2k} = t_{2k}$ and $\widehat{q}_{2k+1} = t_{2k+1}$ satisfy the required inequality.

If $b_{2k+2} = 0$ and $t_{2k} > q_{2k} + q_{2k+1}$ then

$$\widehat{q}_{2k} = \widehat{q}_{2k-1} - q_{2k} + (a_{2k+1} - b_{2k+1})q_{2k}$$
$$\leq \widehat{q}_{2k-1} + a_{2k+1}q_{2k} \leq (q_{2k-1} + q_{2k}) + q_{2k+1} - q_{2k-1}$$
$$= q_{2k} + q_{2k+1}$$

and

$$\widehat{q}_{2k+1} = t_{2k+1} - q_{2k+2} = t_{2k} - q_{2k} = \widehat{q}_{2k}$$
$$\leq q_{2k} + q_{2k+1} < q_{2k+1} + q_{2k+2}. \qquad \square$$

Lemma 7.10 *For $n \geq 0$ set*

$$(\widetilde{p}_n, \widetilde{q}_n) = \begin{cases} (M_n + p_n, N_n + q_n) & \text{if } N_n \leq q_{n+1}, \\ (M_{n+1} + p_{n+1}, N_{n+1} + q_{n+1} = N_n) & \text{if } N_n > q_{n+1}, \end{cases}$$

where N_n denotes the smallest positive solution to (7.2) and

$$M_n = \begin{cases} \lfloor N_n\alpha + \beta \rfloor & \text{if } n \text{ odd}, \\ \lfloor N_n\alpha + \beta \rfloor + 1 & \text{if } n \text{ even}. \end{cases}$$

Then $\widetilde{p}_n = \widehat{p}_n$ and $\widetilde{q}_n = \widehat{q}_n$ for all $n \geq 2m + 1$.

We will not reproduce the proof of this statement from [23], which requires a careful verification of several cases.

Finally, we note that the sequences resulting from the extended continued fraction algorithm have various other continued-fraction-like properties, which are too far afield to discuss here in detail. We introduce just one:

EXERCISE 7.11 (cf. Section 2.9) Show that the number α is a quadratic irrational and β is of the form $u\alpha + v$ for some $u, v \in \mathbb{Q}$ iff both the sequences $\{a_n\}_n$ and $\{b_n\}_n$ are periodic.

7.2 Lambert series expansions of generating functions

The functions below depend analytically on variables z and w that satisfy $|z| \leq 1$, $|w| \leq 1$ and $|zw| < 1$. These hypotheses ensure the convergence of the representing series.

As before, we assume that α and β are two real numbers in the interval $[0, 1)$, with α irrational. Furthermore, we assume that $q\alpha + \beta$ is never an integer for $q \geq 1$; this technical restriction could be dropped in the final result by analyticity.

On the basis of the properties of our extended continued fraction algorithm from Section 7.1, we are in a position to construct Lambert series expansions for the following two-variable generating functions:

$$G(z, w) = G_{\alpha,\beta}(z, w) = \sum_{q=1}^{\infty} z^q w^{\lfloor q\alpha + \beta \rfloor}$$

and

$$F(z, w) = F_{\alpha,\beta}(z, w) = \sum_{q=1}^{\infty} z^q \sum_{p=1}^{\lfloor q\alpha + \beta \rfloor} w^p.$$

EXERCISE 7.12 Prove the duality relation

$$F(z, w) = \frac{zw}{(1 - z)(1 - w)} - \frac{w}{1 - w} G(z, w).$$

In what follows we will use the following rational approximations to our generating functions:

$$G_n(z, w) = G_{n;\alpha,\beta}(z, w) = \frac{1}{1 - z^{q_n} w^{p_n}} \sum_{q=1}^{q_n} z^q w^{\lfloor q\alpha + \beta \rfloor},$$

$$F_n(z, w) = F_{n;\alpha,\beta}(z, w) = \frac{zw}{(1 - z)(1 - w)} - \frac{w}{1 - w} G_n(z, w),$$

where $\{p_n\}_n$ and $\{q_n\}_n$ are the numerators and denominators of the continued fraction of α from the (extended) algorithm. Also, introduce

$$P_n = \sum_{q=1}^{q_n} z^q w^{\lfloor q\alpha+\beta \rfloor} \qquad \text{and} \qquad Q_n = 1 - z^{q_n} w^{p_n},$$

and use all the notation from the previous section.

First consider the functional approximations

$$Q_n G - P_n = \sum_{q=q_n+1}^{\infty} z^q w^{\lfloor q\alpha+\beta \rfloor} - \sum_{q=1}^{\infty} z^{q+q_n} w^{\lfloor q\alpha+\beta+p_n \rfloor}$$

$$= z^{q_n} w^{p_n} \sum_{q=1}^{\infty} z^q \big(w^{\lfloor q\alpha+\beta+\delta_n \rfloor} - w^{\lfloor q\alpha+\beta \rfloor} \big),$$

where $\delta_n = q_n \alpha - p_n$.

It follows from Lemma 7.4 that there exists at most one integer N, $1 \le N \le q_{n+1}$, satisfying (7.2); the smallest such integer N_n may therefore exceed q_{n+1}. Taking the corresponding $M_n = \lfloor N_n \alpha + \beta \rfloor + (1 + (-1)^n)/2$, we get on the one hand

$$Q_n G - P_n = (-1)^{n+1} \frac{1-w}{w} z^{q_n+N_n} w^{p_n+M_n} + O(z^{q_n+q_{n+1}+1} w^{p_n+p_{n+1}-1}). \qquad (7.3)$$

On the other hand,

$$Q_n P_{n+1} - Q_{n+1} P_n = (-1)^{n+1} \frac{1-w}{w} \big(z^{q_n+N_n} w^{p_n+M_n}$$

$$+ z^{q_{n+1}+N_{n+1}} w^{p_{n+1}+M_{n+1}} + O(z^{q_n+q_{n+1}+1}) \big). \qquad (7.4)$$

The left-hand side is a polynomial of degree $q_n + q_{n+1}$ in z, and Lemma 7.5 shows that at most one of the two terms on the right-hand side has a degree that low. Since the left-hand side does not vanish identically, we deduce that

$$\frac{P_{n+1}}{Q_{n+1}} - \frac{P_n}{Q_n} = (-1)^{n+1} \frac{1-w}{w} \frac{z^{\tilde{q}_n} w^{\tilde{p}_n}}{(1 - z^{q_n} w^{p_n})(1 - z^{q_{n+1}} w^{p_{n+1}})}, \qquad (7.5)$$

where the sequences $(\tilde{p}_n)_n$ and $(\tilde{q}_n)_n$ are defined in Lemma 7.10. The quotients P_n/Q_n converge to G by (7.3), so we have for each n

$$G(z,w) = G_n(z,w) + \frac{1-w}{w} \sum_{k=n}^{\infty} \frac{(-1)^{k+1} z^{\tilde{q}_k} w^{\tilde{p}_k}}{(1 - z^{q_k} w^{p_k})(1 - z^{q_{k+1}} w^{p_{k+1}})}. \qquad (7.6)$$

From (7.6) and Exercise 7.12 we also have

$$F(z,w) = F_n(z,w) + \sum_{k=n}^{\infty} \frac{(-1)^k z^{\tilde{q}_k} w^{\tilde{p}_k}}{(1 - z^{q_k} w^{p_k})(1 - z^{q_{k+1}} w^{p_{k+1}})}.$$

The particular instance $n = 0$ of this identity is the case of most interest to us.

Lemma 7.13 *We have*

$$F(z, w) = \sum_{q=1}^{\infty} z^q \sum_{p=1}^{\lfloor q\alpha+\beta \rfloor} w^p = \sum_{n=0}^{\infty} \frac{(-1)^n z^{\widetilde{q}_n} w^{\widetilde{p}_n}}{(1 - z^{q_n} w^{p_n})(1 - z^{q_{n+1}} w^{p_{n+1}})}. \qquad (7.7)$$

On a different route, using the assumption that $q\alpha + \beta$ is never an integer, we can derive the basic functional equation

$$F_{\alpha,\beta}(z, w) + F_{1/\alpha,-\beta/\alpha}(w, z) = \frac{zw}{(1 - z)(1 - w)}, \qquad (7.8)$$

since either $1 \le p \le \lfloor q\alpha + \beta \rfloor$ or $1 \le q \le \lfloor m/\alpha - \beta/\alpha \rfloor$ and not both. In the notation of the extended continued fraction algorithm, the relation (7.8) can be stated as

$$F_{\alpha_0,\beta_0}(z, w) + F_{\alpha_1,-\beta_1}(z, w) = \frac{zw}{(1 - z)(1 - w)}.$$

Letting $z_0 = z$ and $w_0 = w$, we can write more generally

$$F_{\alpha_0,\beta_0}(z_0, w_0) + z_0^{-b_1} F_{\alpha_1,-\beta_1}(z_1, w_1) = z_0^{-b_1} \frac{z_1 w_1 D_1}{(1 - z_1)(1 - w_1)}, \qquad (7.9)$$

where $z_1 = z_0^{a_1} w_0$, $w_1 = z_0$ and $D_1 = z_1^{b_2} + (1 - z_1^{b_2})/w_1$.

Similarly (but in a simpler way since $\lfloor q\alpha_2 + \beta_2 \rfloor$ is always positive while, for $q \le b_2$, $\lfloor q\alpha_1 - \beta_1 \rfloor$ is not), we obtain

$$F_{\alpha_1,-\beta_1}(z_1, w_1) + z_1^{b_2} F_{\alpha_2,\beta_2}(z_2, w_2) = z_1^{b_2} \frac{z_2 w_2}{(1 - z_2)(1 - w_2)}, \qquad (7.10)$$

where $z_2 = z_1^{a_2} w_1$ and $w_2 = z_1$.

Combining (7.9) and (7.10) with indices shifted by any even number gives

$$F_{\alpha_{2k},\beta_{2k}}(z_{2k}, w_{2k}) - z_{2k}^{-b_{2k+1}} z_{2k+1}^{b_{2k+2}} F_{\alpha_{2k+2},\beta_{2k+2}}(z_{2k+2}, w_{2k+2})$$

$$= z_{2k}^{-b_{2k+1}} z_{2k+1}^{b_{2k+2}} \left(\frac{z_{2k+1} w_{2k+1}}{(1 - z_{2k+1})(1 - w_{2k+1})} - \frac{z_{2k+2} w_{2k+2}}{(1 - z_{2k+2})(1 - w_{2k+2})} \right)$$

$$+ \frac{1 - z_{2k+1}^{b_{2k+2}}}{1 - z_{2k+1}} \frac{z_{2k+2}}{1 - z_{2k+2}}, \qquad (7.11)$$

where $z_{n+1} = z_n^{a_n} w_n$, $w_{n+1} = z_n$ and the quantities α_n, β_n, a_n and b_n are generated by the algorithm. The right-hand side of (7.11) can be written as

$$\Delta_{2k} = \frac{z_{2k+1} z_{2k}^{-b_{2k+1}}}{(1 - z_{2k+1})(1 - z_{2k})} - \frac{z_{2k+1}^{b_{2k+2}+1} z_{2k}^{-b_{2k+1}}}{(1 - z_{2k+2})(1 - z_{2k+1})},$$

and we also have

$$\Delta_{2k} = \frac{z_{2k+1} z_{2k} z_{2k}^{-b_{2k+1}}}{(1 - z_{2k+1})(1 - z_{2k})} - \frac{z_{2k+2} z_{2k+1} z_{2k}^{-b_{2k+1}}}{(1 - z_{2k+2})(1 - z_{2k+1})} \qquad \text{when } b_{2k+2} = 0.$$

Iterating the relation (7.11) n times, we obtain

$$F_{\alpha,\beta}(z,w) - x_{2n}F_{\alpha_{2n},\beta_{2n}}(z_{2n},w_{2n}) = \sum_{k=0}^{n-1} x_{2k}\Delta_{2k}, \qquad (7.12)$$

where

$$x_n = \prod_{j=0}^{n-1} z_j^{(-1)^{j+1}b_{j+1}} = x_{n-1}z_{n-1}^{(-1)^n b_n}.$$

To make formula (7.12) explicit, we need to compute z_n and x_n. Inductively, we first find that $z_n = z^{q_n}w^{p_n}$ and then that

$$x_n = z^{t_{n-1}-q_{n-1}-q_n}w^{s_{n-1}-p_{n-1}-p_n}, \qquad n = 0,1,2,\ldots,$$

where, again, the integers s_n and t_n are constructed in the algorithm. With some manipulation we find that

$$x_{2k}\Delta_{2k} = \frac{z^{t_{2k}-q_{2k}}w^{s_{2k}-p_{2k}}}{(1-z_{2k+1})(1-z_{2k})} - \frac{z^{t_{2k+1}-q_{2k+2}}w^{s_{2k+1}-p_{2k+2}}}{(1-z_{2k+2})(1-z_{2k+1})}.$$

Furthermore, $|F_{\alpha_{2n},\beta_{2n}}(z_{2n},w_{2n})| \le |z_{2n}w_{2n}|$ and

$$x_{2n}z_{2n}w_{2n} = z^{t_{2n-1}}w^{s_{2n-1}} \to 0$$

because $t_{2n-1} \to \infty$ as $n \to \infty$ by Lemma 7.7. Therefore, the limit of (7.12) as $n \to \infty$ assumes the form

$$F_{\alpha,\beta}(z,w) = \sum_{n=0}^{\infty}\left(\frac{z^{t_{2n}-q_{2n}}w^{s_{2n}-p_{2n}}}{(1-z_{2n+1})(1-z_{2n})} - \frac{z^{t_{2n+1}-q_{2n+2}}w^{s_{2n+1}-p_{2n+2}}}{(1-z_{2n+2})(1-z_{2n+1})}\right), \qquad (7.13)$$

and the summand can be given by

$$\frac{z^{t_{2n}}w^{s_{2n}}}{(1-z_{2n+1})(1-z_{2n})} - \frac{z^{t_{2n+1}}w^{s_{2n+1}}}{(1-z_{2n+2})(1-z_{2n+1})} \qquad \text{when } b_{2n+2} = 0.$$

Recalling the definitions of $\widehat{p}_n$ and $\widehat{q}_n$ from the extended continued fraction algorithm and using the estimates $q_n \le \widehat{q}_n \le q_n + q_{n+1}$ of Lemmas 7.8 and 7.9, we can restate (7.13) as follows.

Lemma 7.14 *We have*

$$F(z,w) = \sum_{n=0}^{\infty} \frac{(-1)^n z^{\widehat{q}_n}w^{\widehat{p}_n}}{(1-z^{q_n}w^{p_n})(1-z^{q_{n+1}}w^{p_{n+1}})}. \qquad (7.14)$$

Lemmas 7.13 and 7.14 produce two (similar) representations for the same function $F(z, w)$, especially if we recall from Lemma 7.10 that eventually $\widetilde{p}_n = \widehat{p}_n$ and $\widetilde{q}_n = \widehat{q}_n$. We will require truncations of the following infinite series:

$$\widetilde{F}_n(z, w) = \sum_{n=0}^{\infty} \frac{(-1)^n z^{\widetilde{q}_n} w^{\widetilde{p}_n}}{(1 - z^{q_n} w^{p_n})(1 - z^{q_{n+1}} w^{p_{n+1}})}$$

and

$$\widehat{F}_n(z, w) = \sum_{n=0}^{\infty} \frac{(-1)^n z^{\widehat{q}_n} w^{\widehat{p}_n}}{(1 - z^{q_n} w^{p_n})(1 - z^{q_{n+1}} w^{p_{n+1}})}.$$

The next result is then straightforward from these definitions and Lemmas 7.10–7.14.

Lemma 7.15   *With the earlier notation* $Q_n = 1 - z^{q_n} w^{p_n}$, *we have for* $n > l$

$$\widetilde{F}_n - \widetilde{F}_l = \sum_{k=l}^{n-1} \frac{(-1)^k z^{\widetilde{q}_k} w^{\widetilde{p}_k}}{Q_k Q_{k+1}} \qquad and \qquad \widehat{F}_n - \widehat{F}_l = \sum_{k=l}^{n-1} \frac{(-1)^k z^{\widehat{q}_k} w^{\widehat{p}_k}}{Q_k Q_{k+1}}.$$

Furthermore, if $a_{2m+1} > b_{2m+1}$ *then for all* $n, l \geq 2m + 1$

$$\widetilde{F}_n - \widetilde{F}_l = \widehat{F}_n - \widehat{F}_l \qquad and \qquad \widetilde{F}_n = \widehat{F}_n.$$

Theorem 7.16   *Assume that* α, β *are real numbers from the interval* $[0, 1)$, α *is irrational and* $q\alpha + \beta$ *is never an integer for* $q > 0$. *The integer* m *is defined as the smallest positive integer for which* $a_{2m+1} > b_{2m+1}$.

(i) *For* $n \geq 2m + 1$,

$$\sum_{k=0}^{n-1} \frac{(-1)^k z^{\widehat{q}_k} w^{\widehat{p}_k}}{(1 - z^{q_k} w^{p_k})(1 - z^{q_{k+1}} w^{p_{k+1}})}$$

$$= \frac{zw}{(1-z)(1-w)} - \frac{w}{1-w} \frac{1}{1 - z^{q_n} w^{p_n}} \sum_{q=1}^{q_n} z^q w^{\lfloor q\alpha+\beta \rfloor}$$

with $\widehat{p}_n$ *and* $\widehat{q}_n$ *defined in the extended continued fraction algorithm. In particular, if* $m = 0$, *which happens if* $\alpha + \beta < 1$ *or* $\beta < \alpha$ *and therefore if* $0 \leq \beta \leq 1/2$, *then the identity holds for* $n = 1, 2, \ldots$

(ii) *For* $n \geq 1$,

$$\frac{zw}{(1-z)(1-w)} - \frac{w}{1-w} \frac{1}{1 - z^{q_n} w^{p_n}} \sum_{q=1}^{q_n} z^q w^{\lfloor q\alpha+\beta \rfloor} = \frac{\Lambda_n(z, w)}{(1-z)(1 - z^{q_n} w^{p_n})},$$

where the polynomial $\Lambda_n \in \mathbb{Z}[z, w]$ *has degree less than* q_{n+1} *in* z.

Proof Part (i) follows from Lemma 7.15, while in part (ii) we use

$$\Lambda_n(z, w) = wz\left(\frac{1 - w^{\lfloor \alpha+\beta \rfloor}}{1 - w} - z^{q_n} w^{p_n} \frac{1 - w^{\lfloor \beta+\delta_n \rfloor}}{1 - w}\right.$$

$$\left. + \sum_{q=1}^{q_n-1} (\lfloor (q+1)\alpha + \beta \rfloor - \lfloor q\alpha + \beta \rfloor) z^{q-1} w^{\lfloor q\alpha+\beta \rfloor}\right).$$

Note that $\lfloor \alpha + \beta \rfloor = 0$ or 1, while $\lfloor \beta + \delta_n \rfloor = 0$ or -1; the latter is in fact 0 for n even or n large and $\beta > 0$.   □

Taking $w = 1$ in Theorem 7.16 we have for $\beta \geq 0$

$$\sum_{q=1}^{\infty} \lfloor q\alpha + \beta \rfloor z^q = \sum_{n=0}^{\infty} \frac{(-1)^n z^{\widehat{q_n}}}{(1 - z^{q_n})(1 - z^{q_{n+1}})}, \tag{7.15}$$

and the rational-function truncation

$$\sum_{k=0}^{n-1} \frac{(-1)^k z^{\widehat{q_k}}}{(1 - z^{q_k})(1 - z^{q_{k+1}})} = \frac{\Lambda_n(z, 1)}{(1 - z)(1 - z^{q_n})}$$

agrees with the left-hand side of (7.15) to $O(z^{q_n})$.

As a general corollary of Theorem 7.16 we can state the following result.

Theorem 7.17 *Let complex z and w satisfy $|z| \leq 1$, $|w| \leq 1$ and $|zw| < 1$. Assume that α, β are real numbers from the interval $[0, 1)$, α is irrational and $\alpha + \beta < 1$. Then*

$$G_{\alpha,\beta}(z, w) = \sum_{q=1}^{\infty} z^q w^{\lfloor q\alpha+\beta \rfloor}$$

$$= \frac{z}{1 - z} + \frac{1 - w}{w} \sum_{n=0}^{\infty} \frac{(-1)^{n+1} z^{\widehat{q_n}} w^{\widehat{p_n}}}{(1 - z^{q_n} w^{p_n})(1 - z^{q_{n+1}} w^{p_{n+1}})} \tag{7.16}$$

and

$$F_{\alpha,\beta}(z, w) = \sum_{q=1}^{\infty} z^q \sum_{p=1}^{\lfloor q\alpha+\beta \rfloor} w^p = \sum_{n=0}^{\infty} \frac{(-1)^n z^{\widehat{q_n}} w^{\widehat{p_n}}}{(1 - z^{q_n} w^{p_n})(1 - z^{q_{n+1}} w^{p_{n+1}})}.$$

Note that the assumption $\alpha + \beta < 1$ may be dropped by the inclusion of some additional initial terms.

It is not hard to see that, in the special case $\beta = 0$, the extended algorithm of Section 7.1 produces $\widehat{p_n} = p_n + p_{n+1}$ and $\widehat{q_n} = q_n + q_{n+1}$. Theorem 7.17 for this case is due to Mahler [108] and is discussed in more detail in [104]. Furthermore, in the case $\beta = 0$ we can give a continued fraction expansion related to the generating function $G(z, w)$.

Theorem 7.18   *For an irrational number $\alpha > 0$ and complex z and w, $|z| \leq 1$,
$|w| < 1$,*

$$\frac{1-w}{w}\sum_{q=1}^{\infty} z^q w^{\lfloor q\alpha \rfloor} = [0; A_0, A_1, A_2, \dots] = \cfrac{1}{A_0 + \cfrac{1}{A_1 + \cfrac{1}{A_2 + \ddots}}},$$

where

$$A_0 = \frac{z^{-1}w^{-a_0} - 1}{w^{-1} - 1}$$

and

$$A_n = \frac{z^{-q_{n-2}}w^{-p_{n-2}}\left(z^{-a_n q_{n-1}}w^{-a_n p_{n-1}} - 1\right)}{z^{-q_{n-1}}w^{-p_{n-1}} - 1} \quad \text{for } n \geq 1,$$

with a_n the partial quotients and p_n/q_n the convergents of the continued fraction expansion of α.

Proof   Take

$$\widetilde{P}_n = z^{-q_n}w^{-p_n}\sum_{q=1}^{q_n} z^q w^{\lfloor q\alpha \rfloor} \quad \text{and} \quad \widetilde{Q}_n = z^{-q_n}w^{-p_n} - 1.$$

Observe that with $\beta = 0$ we have by (7.5)

$$\frac{\widetilde{P}_{n+1}}{\widetilde{Q}_{n+1}} - \frac{\widetilde{P}_n}{\widetilde{Q}_n} = (-1)^{n+1}\frac{1-w}{w}\frac{1}{\widetilde{Q}_{n+1}\widetilde{Q}_n}$$

and hence

$$\frac{\widetilde{P}_{n+1}}{\widetilde{Q}_{n+1}} - \frac{\widetilde{P}_{n-1}}{\widetilde{Q}_{n-1}} = \frac{1-w}{w}\frac{(-1)^n A_{n+1}}{\widetilde{Q}_{n+1}\widetilde{Q}_{n-1}},$$

which implies the matrix form

$$\begin{pmatrix}\widetilde{P}_{n+1}\\ \widetilde{Q}_{n+1}\end{pmatrix} = A_{n+1}\begin{pmatrix}\widetilde{P}_n\\ \widetilde{Q}_n\end{pmatrix}\begin{pmatrix}\widetilde{P}_{n-1}\\ \widetilde{Q}_{n-1}\end{pmatrix}$$

with $P_{-1} = 0$, $P_0 = 1$, $Q_{-1} = w^{-1} - 1$ and $Q_0 = z^{-1}w^{-a_0} - 1$. It follows from the results in Chapter 2 that the A_n are the partial quotients of a continued fraction, while $\widetilde{P}_n/\widetilde{Q}_n$ are the convergents. It remains to use the convergence

$$\frac{\widetilde{P}_n}{\widetilde{Q}_n} \to \sum_{q=1}^{\infty} z^q w^{\lfloor q\alpha \rfloor} \quad \text{as } n \to \infty. \qquad \square$$

EXERCISE 7.19   Show that if $\alpha = (1 + \sqrt{5})/2$ then

$$\frac{1-w}{w} \sum_{q=1}^{\infty} z^q w^{\lfloor q\alpha \rfloor} = \left[0; \frac{z^{-1}w^{-1}-1}{w^{-1}-1}, z^{-F_0}w^{-F_1}, z^{-F_1}w^{-F_2}, z^{-F_2}w^{-F_3}, \dots \right],$$

where $F_0, F_1, \dots$ are the Fibonacci numbers (from Section 1.3).

EXERCISE 7.20   Let $\beta > 0$ be irrational and define

$$f_\beta(n) = \lfloor \beta(n+1) \rfloor - \lfloor \beta n \rfloor,$$

the so-called *characteristic sequence* associated with β. Show that

$$\sum_{q=1}^{\infty} f_\beta(q) t^{-q} = \sum_{q=1}^{\infty} t^{-\lfloor q/\beta \rfloor} = (t-1) \sum_{q=1}^{\infty} \lfloor q\beta \rfloor t^{-q}.$$

Next, by setting in Theorem 7.18 $w = t^{-1}$, $z = 1$ and $\alpha = 1/\beta$, find the continued fraction expansion of $(t-1) \sum_{q=1}^{\infty} f_\beta(q) t^{-q}$. (This is the celebrated Böhmer–Danilov–Adams–Davison continued fraction, which has been redis-covered multiple times; see [17, 52, 55, 2, 154]. For a generalisation, see [34].)

As a corollary of Theorem 7.17 we can fairly easily establish the irrationality of the values of $G_{\alpha,\beta}(z, w)$, at least in the special case $z = 1/u$ and $w = 1/v$ where u and v are positive integers with $uv > 1$. This follows from noting that the integers

$$\widehat{Q}_n = u^{q_n} v^{p_n} Q_n\!\left(\frac{1}{u}, \frac{1}{v}\right) = u^{q_n} v^{p_n} - 1$$

and

$$\widehat{P}_n = u^{q_n} v^{p_n} P_n\!\left(\frac{1}{u}, \frac{1}{v}\right) = u^{q_n} v^{p_n} \sum_{q=1}^{q_n} u^{-q} v^{-\lfloor q\alpha+\beta \rfloor}$$

satisfy

$$0 < \left| G\!\left(\frac{1}{u}, \frac{1}{v}\right) - \frac{\widehat{P}_n}{\widehat{Q}_n} \right| < \widehat{Q}_n^{-(1+q_{n+1}/q_n)/(2+\delta)}$$

for all n sufficiently large (depending on $\delta > 0$). Because

$$\liminf_{n\to\infty} \frac{q_{n+1}}{q_n} \frac{1+\sqrt{5}}{2} > 1,$$

the former estimate implies the irrationality of $G(1/u, 1/v)$; see Theorem 1.34.

In fact, an analysis based on the powerful method of Mahler [108, 118] al-lows one to demonstrate the transcendence of the values $F_{\alpha,\beta}(z, w)$ or $G_{\alpha,\beta}(z, w)$ for any nonzero *algebraic* z and w; indeed, even the algebraic independence of the values at several algebraic points can be proved; see [85, Theorem 6] for details.

A different application, with a more combinatorial flavour, is the following result originally due to Skolem [155] (also see Fraenkel [61]). The particular case $\beta = 0$ is known as Beatty's theorem (although Beatty was just one of its many rediscoverers after Lord Raleigh).

Theorem 7.21 *Let* $\alpha > 0$ *and* $\alpha' > 0$ *be irrational and* β *and* β' *real such that* $q\alpha + \beta$ *and* $q\alpha' + \beta'$ *are never integers. Then the two sets*

$$\{\lfloor q\alpha + \beta \rfloor\}_{q=1}^{\infty} \quad and \quad \{\lfloor q\alpha' + \beta' \rfloor\}_{q=1}^{\infty}$$

disjointly partition the set of positive integers iff

$$\frac{1}{\alpha} + \frac{1}{\alpha'} = 1 \quad and \quad \frac{\beta}{\alpha} + \frac{\beta'}{\alpha'} = 0.$$

Proof First note that if $\alpha \leq 1$ then the corresponding set $\{\lfloor q\alpha + \beta \rfloor\}_{q=1}^{\infty}$ already contains all integers greater than β. This means that $\alpha > 1$ and $\alpha' > 1$.

The partitioning property is equivalent to

$$\sum_{q=1}^{\infty} z^{\lfloor q\alpha + \beta \rfloor} + \sum_{q=1}^{\infty} z^{\lfloor q\alpha' + \beta' \rfloor} = \sum_{q=1}^{\infty} z^q = \frac{z}{1-z}.$$

Although the series on the left-hand side are easily recognised as $G_{\alpha,\beta}(1,z)$ and $G_{\alpha',\beta'}(1,z)$, we will require the less obvious forms $G_{\alpha-1,\beta}(z,z)$ and $G_{\alpha'-1,\beta'}(z,z)$:

$$G_{\alpha-1,\beta}(z,z) + G_{\alpha'-1,\beta'}(z,z) = \frac{z}{1-z}. \qquad (7.17)$$

Using the duality relation of Exercise 7.12 we can state (7.8),

$$F_{\alpha-1,\beta}(z,w) + F_{1/(\alpha-1),-\beta/(\alpha-1)}(w,z) = \frac{zw}{(1-z)(1-w)},$$

in the form

$$G_{1/(\alpha-1),-\beta/(\alpha-1)}(w,z) + \frac{1-z}{z}\frac{w}{1-w} G_{\alpha-1,\beta}(z,w) = \frac{w}{1-w}.$$

Specialising to $w = z$ and comparing with (7.17) produces

$$G_{1/(\alpha-1),-\beta/(\alpha-1)}(z,z) = G_{\alpha'-1,\beta'}(z,z),$$

which holds iff, for large n, the integer parts $\lfloor q/(\alpha-1) - \beta/(\alpha-1) \rfloor$ and $\lfloor q(\alpha'-1) + \beta' \rfloor$ coincide. By the uniformity of the distribution it follows that $\alpha' - 1 = 1/(\alpha-1)$ and $\beta' = -\beta/(\alpha-1)$, which are equivalent to the desired relations. $\qquad \square$

7.3 High-precision fraud

When $\beta = 0$ and irrational α is close to an integer, the values (which are irrational and even transcendental by [104]) of the generating series

$$F_{\alpha,0}(z, 1) = \sum_{q=1}^{\infty} \lfloor q\alpha \rfloor z^q$$

are very close to rational numbers. This is a fantastic source of 'fraudulent' evaluations [22] such as

$$\sum_{q=1}^{\infty} \frac{\lfloor q \tanh(\pi) \rfloor}{10^q} \stackrel{?}{=} \frac{1}{81},$$

which is wrong, although valid to 268 decimal places. Note that $0.99 < \tanh(\pi) < 1$ and $\lfloor q \tanh(\pi) \rfloor$ is equal to $q - 1$ for many initial q; precisely, for $q = 1, \ldots, 268$ because

$$\tanh(\pi) = [0; 1, 267, 4, 14, 1, 2, 1, 2, 2, 1, 2, 3, 8, 3, 1, \ldots].$$

Since

$$\sum_{q=1}^{\infty} \frac{q-1}{10^q} = \frac{1}{81},$$

this explains the evaluation. As we know from the results of the previous section, it is not coincidental that the integer 267 appears as the second partial quotient: we have $q_0 = 1, q_1 = 1, q_2 = 268, q_3 = 1073, \ldots$, so that

$$\sum_{q=1}^{\infty} \lfloor q \tanh(\pi) \rfloor z^q = \frac{z^2}{(1-z)^2} - \frac{z^{269}}{(1-z)(1-z^{268})} + \cdots$$

by Theorem 7.17 and

$$\frac{1}{81} - 2 \times 10^{-269} \le \sum_{q=1}^{\infty} \frac{\lfloor q \tanh(\pi) \rfloor}{10^q} \le \frac{1}{81} + 2 \times 10^{-269}.$$

With another favourite *transcendental* number,

$$\alpha = e^{\pi\sqrt{163/9}} = [640\,320; 1\,653\,264\,929, 30, 1, 321, \ldots],$$

we get the incorrect evaluation

$$\sum_{q=1}^{\infty} \frac{\lfloor q e^{\pi\sqrt{163/9}} \rfloor}{2^q} \stackrel{?}{=} 1\,280\,640,$$

which is correct to at least half a billion digits! (The proof of its incorrectness exploits the irrationality of the left-hand side.)

In our final example we take $\alpha = \log_{10}(2)$, so that $\lfloor q\alpha \rfloor + 1$ counts the number of decimal digits of 2^q. Then $q_0 = 1$, $q_1 = 3$, $q_2 = 10$, $q_3 = 93$, ..., and as a result the transcendental number

$$\sum_{q=1}^{\infty} \frac{\lfloor q \log_{10}(2) \rfloor}{2^q}$$

is equal to $146/1023$ to 30 decimal digits. Interestingly, if $\Sigma_{\text{even}}(n)$ (resp. $\Sigma_{\text{odd}}(n)$) counts the number of even (resp. odd) decimal digits of n then

$$\sum_{q=1}^{\infty} \frac{\Sigma_{\text{odd}}(2^q)}{2^q}$$

is perfectly rational, while

$$\sum_{q=1}^{\infty} \frac{\Sigma_{\text{even}}(2^q)}{2^q} = \sum_{q=1}^{\infty} \frac{\lfloor q \log_{10}(2) \rfloor + 1}{2^q} - \sum_{q=1}^{\infty} \frac{\Sigma_{\text{odd}}(2^q)}{2^q}$$

is transcendental. Again, the transcendence result follows from [104], while an evaluation for the sum with $\Sigma_{\text{odd}}(2^q)$ is the subject of the following exercise.

EXERCISE 7.22 ([22, Theorem 1.7]) If $\Sigma_{\text{odd}}(n)$ counts the number of odd decimal digits of n then show that

$$\sum_{q=1}^{\infty} \frac{\Sigma_{\text{odd}}(2^q)}{2^q} = \frac{1}{9}.$$

Notes

The Lambert series representation for the generating function $F_{\alpha,\beta}(z, w)$ given in Theorem 7.17 is not unique. Correcting certain details of arguments given by Nishioka, Shiokawa and Tamura [119], Komatsu constructed a different expansion of this type in [85] and used it to derive various arithmetical results for values of the function. He also verified the coincidence of both Lambert series by direct means.

In [87] Komatsu used the extended continued fraction algorithm of Section 7.1 to estimate and compute the *inhomogeneous approximation constant*

$$M(\alpha, \beta) = \liminf_{|q| \to \infty} |q| \, \|q\alpha - \beta\|,$$

where the norm $\| \cdot \|$ measures the distance to the nearest integer, α is irrational

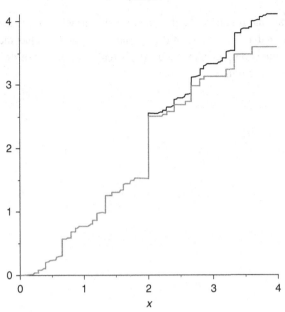

Figure 7.1 The summand in $K_{1/2}(7/16, x)$ plotted to 10 terms (lower curve), 100 and 1000 terms (upper curve).

and $q\alpha - \beta$ is assumed not to be an integer for all $q \in \mathbb{Z}$. In particular, Komatsu showed [86, 87] that

$$M\left(\frac{\sqrt{ab(ab+4)} - ab}{2a}, \frac{1}{b}\right) = \frac{a}{b^2\sqrt{ab(ab+4)}}$$

for integers $a \geq 1$ and $b \geq 2$, and

$$M\left(e, \frac{1}{k}\right) = \frac{1}{2k^2} \qquad \text{for } k = 2, 3, 4.$$

The question of whether the latter result holds for all $k \geq 2$ remains open.

There is an incomplete discussion about generalisations of Theorem 7.21 involving partitioning the positive integers by more than two sets of the form $S(\alpha, \beta) = \{\lfloor q\alpha + \beta\rfloor\}_{n=1}^{\infty}$. The particular case of pairwise distinct rates α was first discussed by Fraenkel [62], who made an explicit conjecture about their choice. Graham [67] then showed that under Fraenkel's hypothesis the rates should be rational. The latest progress towards a full proof of Fraenkel's conjecture is reported in [11, 160, 161].

As noted in [122] the special case of (7.16) given by $K_\gamma(z, x) = \sum_{n=1}^{\infty} z^{\lfloor n/x+\gamma\rfloor}$ (see Figure 7.1) has several exotic properties. It is strictly increasing in x and, if

$\gamma = 0$, is a rational function of z iff x is rational. Further, $K_\gamma(z, x)$ is continuous only at irrational x. The image of $(0, \infty)$ under $x \mapsto K_\gamma(z, x)$ has measure zero as a subset of the complex plane and – if z is real – as a subset of the reals. This is illustrated in the figure.

8

The Erdős–Moser equation

In this chapter we will be interested in nontrivial integer solutions of the equation

$$1^k + 2^k + \cdots + (m-2)^k + (m-1)^k = m^k, \qquad (8.1)$$

for $k \geq 2$. The conjecture that such solutions do not exist was formulated around 1950 by Erdős in a letter to Moser. For $k = 1$, one has clearly the solution $1 + 2 = 3$ (and no others). The equation (8.1) seems to be the sole example of an exponential Diophantine equation in just two unknowns for which even the *finiteness* of the solution set is not yet established.

8.1 Arithmetic and analysis of the equation

Moser [116] showed in 1953 that if (m, k) is a solution of (8.1) then $m > 10^{10^6}$ and k is even. This was a remarkable achievement in the pre-computer era, and the argument was based on an elementary arithmetic observation (see also [114]): the four sums over primes

$$\left(\sum_{p|m-1} \frac{1}{p} \right) + \frac{1}{m-1}, \qquad \left(\sum_{p|m+1} \frac{1}{p} \right) + \frac{2}{m+1},$$

$$\left(\sum_{p|2m-1} \frac{1}{p} \right) + \frac{2}{2m-1} \quad \text{and} \quad \left(\sum_{p|2m+1} \frac{1}{p} \right) + \frac{4}{2m+1}$$

are integers, which are clearly positive.

Introducing $M = (m-1)(m+1)(2m-1)(2m+1)/12$ as the square-free part

147

of the product of $m \pm 1$ and $2m \pm 1$, and combining the four sums together, we see that

$$\frac{1}{2} + \frac{1}{3} + \left(\sum_{p|M} \frac{1}{p}\right) + \frac{1}{m-1} + \frac{2}{m+1} + \frac{2}{2m-1} + \frac{4}{2m+1}$$

is an integer which is at least 4. At the same time, the sum of reciprocals of the first $4\,990\,906$ primes is less than $4 - 1/2 - 1/3 - 10^{-9}$, and this implies that M is bounded from below by the product of these primes. This leads one to Moser's lower bound on m, which can be slightly sharpened [38] to $m > 1.485 \times 10^{9\,321\,155}$ by an explicit computation. A disadvantage of Moser's proof is that one cannot go further by extending the argument [114].

In a different direction Moree *et al.* [115], using certain divisibility properties of the Bernoulli numbers and polynomials, showed that k has a large divisor.

REMARK 8.1   As discovered by Jakob Bernoulli (1654–1705), the relation of such sums to Bernoulli polynomials is specified by

$$\sum_{j=1}^{m-1} j^{n-1} = \frac{B_n(m) - B_n}{n},$$

where

$$B_n(x) = \sum_{k=0}^{n} \binom{n}{k} B_k x^{n-k},$$

with the Bernoulli numbers B_n generated by

$$\frac{t}{e^t - 1} = \sum_{n=0}^{\infty} B_n \frac{t^n}{n!},$$

or by

$$\sum_{k=0}^{n} \binom{n}{k} \frac{B_k}{n+2-k} = \frac{B_{n+1}}{n+1}.$$

Recall that the only nonzero odd Bernoulli number is $B_1 = -1/2$, while

$$B_2 = \frac{1}{6}, \quad B_4 = -\frac{1}{30}, \quad B_6 = \frac{1}{42}, \quad B_8 = -\frac{1}{30}, \quad B_{10} = \frac{5}{66}, \quad \cdots$$

Figure 8.1 shows the Bernoulli polynomials for $n \le 6$. Armed with only the Bernoulli numbers listed above, Bernoulli wrote:

With the help of this table, it took me less than half of a quarter of an hour to find that the tenth powers of the first 1000 numbers being added together will yield the sum 91 409 924 241 424 243 424 241 924 242 500.

It is noted in MacTutor[1] that Bernoulli's pleasure in his discovery was great. This may be considered the first triumph of computational number theory.

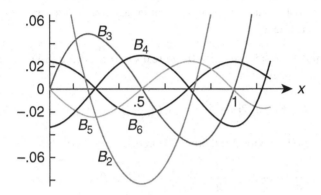

Figure 8.1 The first few Bernoulli polynomials.

EXERCISE 8.2 Explain the palindrome within Bernoulli's computation of $\sum_{j=1}^{1000} j^{10}$. Note that

$$\sum_{j=1}^{10000} j^{10} = 9\,095\,909\,924\,242\,414\,242\,424\,342\,424\,241\,924\,242\,425\,000.$$

The approach of [115] was extended later by Kellner [79], and the final result of [79, 115] can be summarised as follows. The number

$$N^* = 2^8 \times 3^5 \times 5^4 \times 7^3 \times 11^2 \times 13^2 \times 17^2 \times 19^2 \times \prod_{23 \le p \le 997} p \qquad (8.2)$$
$$> 5.7462 \times 10^{427}$$

divides the number k from any nontrivial solution (m, k) of (8.1). This gives the lower bound $k \ge N$ on k and the same bound for m, which, of course, is very modest compared with Moser's, because estimating the sum $\Sigma_k(m) = 1^k + 2^k + \cdots + m^k$ by suitable integrals (as we did in our proof of Lemma 2.59), one easily obtains

$$k + 2 < m < 2k \qquad \text{for } k \ge 4. \qquad (8.3)$$

---

[1] See http://en.wikipedia.org/wiki/Bernoulli_number.

Some further references and a historical account of the Erdős–Moser equation can be found in the book [70, D7].

EXERCISE 8.3 ([113, Proposition 9])   Show that if $p \mid m$ then $p - 1 \nmid k$.

In a more analytic fashion, the dependence of k on m (or of m on k) can be determined more explicitly. Indeed, dividing both sides of (8.1) by m^k and using (8.3) one sees that, for every integer $m \geq 2$, (8.1) has precisely one *real* solution k. With some work one sees that in fact $k/m \to \log 2$ as $m \to \infty$, and even more is true.

Theorem 8.4   *For integer $m > 0$ and real $k > 0$ satisfying* (8.1), *we have the asymptotic expansion*

$$k = \log 2\left(m - \frac{3}{2} - \frac{c_1}{m} + O\left(\frac{1}{m^2}\right)\right) \qquad as\ m \to \infty, \qquad (8.4)$$

with $c_1 = 25/12 - 3\log 2 \approx 0.00389$. Moreover, if $m > 10^9$ then

$$\frac{k}{m} = \log 2\left(1 - \frac{3}{2m} - \frac{C_m}{m^2}\right), \qquad where\ 0 < C_m < 0.004. \qquad (8.5)$$

Though the proof of this statement is given in [65], the analytical approach and also the idea of relating the quotient of k and m to the continued fraction of $\log 2$ go back to the 1976 preprint of Best and te Riele [15], in which they attempted to solve the related conjecture of Erdős [57] that there are infinitely many pairs (m, k) such that $\Sigma_k(m) \geq m^k$ and $2(m - 1)^k < m^k$. We do not reproduce here the details (which are quite technical) from [65] of proving Theorem 8.4, but our discussion in Section 8.2 should provide the reader with sufficient (and elementary) hints to establish the result.

Corollary 8.5   *If (m, k) is a solution of* (8.1) *with $k \geq 2$ then $2k/(2m - 3)$ is a convergent p_n/q_n of $\log 2$ with n even.*

Corollary 8.6   *The number of solutions $m \leq x$ of* (8.1), *as x tends to infinity, is $O(\log x)$.*

Corollary 8.5 follows from the bounds on $2k/(2m-3)$ derived from (8.5) and Legendre's theorem (Lemma 2.33), while Corollary 8.6 is an immediate consequence of the exponential growth of p_n as a function of n and Corollary 8.5.

The principal conclusion here is that the convergents p_n/q_n of $\log 2$ have no particular reason to satisfy also the condition $N^* \mid p_n$, where the monster N^* is defined in (8.2). A symbiosis of the arithmetic and analytical information above can be summarised as follows.

Theorem 8.7 *Let $N \geq 1$ be an arbitrary integer. Let*

$$\frac{\log 2}{2N} = [a_0; a_1, a_2, \ldots]$$

be the continued fraction for $(\log 2)/(2N)$, with $p_n/q_n = [a_0; a_1, \ldots, a_n]$ its nth convergent.

Suppose that the integer pair (m, k) with $k \geq 2$ satisfies (8.1) with $N \mid k$. Let $n = n(N)$ be the smallest index such that

(i) *n is even;*

(ii) *$a_{n+1} \geq 180N - 2$; and*

(iii) *$(q_n, 6) = 1$.*

Then $m > q_n/2$.

REMARK 8.8 Corollary 8.5 naturally leads one to an investigation of the common factors of k and $2m - 3$; the corresponding result allows one to add another arithmetic condition about the primes p dividing q_n, such that $p - 1 \mid N$. The resulting criterion, which is stronger than Theorem 8.7, is given in [65, Theorem 2].

Computational strategies for the continued fraction expansion of a number such as $(\log 2)/(2N)$ are discussed below in Section 8.3. Lochs' theorem (see Exercise 3.9) implies that for a 'typical' number, knowing it accurately up to d decimal digits gives a way of computing about $0.97d$ of its partial quotients.

Although it might seem a hopeless problem to prove anything about the expected value of $\log q_{n(N)}$ in Theorem 8.7, one can use the metric theory of continued fractions presented in Chapter 3 and the hypothesis on the generosity of $\log 2$ to produce some reasonable heuristics about this quantity. This is done in [65], and the heuristic conclusion there is that one might be able to show $m > 10^{10^{400}}$ using $N = N^*$ from (8.2). Realistically, with the current computing technology, one could hope to verify (with a lot of effort!) only that $m > 10^{10^{12}}$. The result given in [65] uses $N = 2^8 \times 3^5 \times 5^3$ in Theorem 8.7 and is as follows.

Theorem 8.9 (Potential counter-examples) *If an integer pair (m, k) with $k \geq 2$ satisfies (8.1) then*

$$m > 2.7139 \times 10^{1\,667\,658\,416}.$$

Details of the derivation of the theorem are postponed until Section 8.4, while in the next section we discuss the asymptotics in Theorem 8.4 and the proof of Theorem 8.7.

8.2 Asymptotics and continued fraction for $\log 2$

Lemma 8.10 *For any real $k > 0$ we have*

$$(1-y)^k = e^{-ky}\left(1 - \frac{k}{2}y^2 - \frac{k}{3}y^3 + \frac{k(k-2)}{8}y^4 + \frac{k(5k-6)}{30}y^5 + \cdots\right)$$

$$= e^{-ky}\sum_{n=0}^{\infty} g_n(k)y^n \qquad \text{as } y \to 0, \tag{8.6}$$

where the coefficients $g_n(k)$ are polynomials in k that satisfy

$$g_0(k) = 1, \quad g_1(k) = 0$$

and

$$\deg_k g_n(k) = \left[\frac{n}{2}\right], \quad g_n(0) = 0 \qquad \text{for } n \geq 2.$$

Proof This follows from raising the series

$$(1-y)e^y = 1 - \sum_{n=2}^{\infty} \frac{n-1}{n!}y^n$$

$$= 1 - \frac{y^2}{2} - \frac{y^3}{3} - \frac{y^4}{8} - \frac{y^5}{30} - \cdots$$

to the power k. □

Unfortunately, estimates coming from the classical forms for the remainder are not sufficient to derive the sufficiently sharp dependence on k (rather than y) required for proving Theorem 8.4.

Lemma 8.11 *For positive real t consider the generalised Erdős–Moser equation*

$$1^k + 2^k + \cdots + (m-1)^k = tm^k, \tag{8.7}$$

which defines k as a function of m. Then the following asymptotic expansion is valid as $m \to \infty$:

$$k = cm + c_0 + \frac{c_1}{m} + \frac{c_2}{m^2} + \cdots, \tag{8.8}$$

where $c = \log(1 + 1/t)$ and c_n are polynomials in t and c of degree $2n + 1$ and $n + 1$, respectively.

Sketch of proof   Using Lemma 8.10 we can write (8.7) in the form

$$t = \sum_{j=1}^{m-1}\left(1 - \frac{j}{m}\right)^k = \sum_{j=1}^{m-1} e^{-kj/m} \sum_{n=0}^{\infty} g_n(k)\left(\frac{j}{m}\right)^n$$

$$= \sum_{n=0}^{\infty} \frac{g_n(k)}{m^n} \sum_{j=1}^{m-1} j^n e^{-jk/m}.$$

Now, since $\sum_{j=m}^{\infty} j^n e^{-jk/m} = O(m^n e^{-k})$, we have

$$t \sim \sum_{n=0}^{\infty} \frac{g_n(k)}{m^n} \sum_{j=1}^{\infty} j^n e^{-jk/m} = \sum_{n=0}^{\infty} \frac{g_n(k)}{m^n}\left(\left(z\frac{\mathrm{d}}{\mathrm{d}z}\right)^n \frac{z}{1-z}\right)\Bigg|_{z=e^{-k/m}}$$

$$= \sum_{n=0}^{\infty} \frac{g_n(k)}{m^n}(-1)^n\left(\left(z\frac{\mathrm{d}}{\mathrm{d}z}\right)^n \frac{1}{z-1}\right)\Bigg|_{z=e^{k/m}}.$$

Hence, with $\lambda = k/m$ and $x = 1/m$, we have

$$t = \sum_{n=0}^{\infty} g_n\left(\frac{\lambda}{x}\right)(-x)^n\left(\left(z\frac{\mathrm{d}}{\mathrm{d}z}\right)^n \frac{1}{z-1}\right)\Bigg|_{z=e^{\lambda}}. \tag{8.9}$$

Searching for λ in the form $\lambda = c + c_0 x + c_1 x^2 + c_2 x^3 + \cdots$, we find successively

$$c = \log\left(1 + \frac{1}{t}\right) = \log\frac{t+1}{t}, \qquad c_1 = -\left(t + \frac{1}{2}\right)c,$$

$$c_2 = \left(t + \frac{1}{2}\right)^3 c^2 - \left(t + \frac{1}{2}\right)^2 c - \frac{1}{4}\left(t + \frac{1}{2}\right)c^2 + \frac{c}{6}$$

and so on.   □

Note that $c_n(-(t + 1)) = (-1)^{n+1}c_n(t)$ for $n = 0, 1, 2, \ldots$; this reflects the equivalence of (8.7) and

$$1^k + 2^k + \cdots + (m - 1)^k + m^k = (t + 1)m^k. \tag{8.10}$$

The asymptotics for the original Erdős–Moser equation follows from specialising to $t = 1$ in Lemma 8.11:

$$k = cm - \frac{3}{2}c - \left(\frac{25}{12}c - 3c^2\right)m^{-1} + \left(-\frac{73}{8}c + \frac{61}{2}c^2 - 25c^3\right)m^{-2}$$

$$+ \left(-\frac{41299}{720}c + \frac{657}{2}c^2 - 598c^3 + \frac{1405}{4}c^4\right)m^{-3} + O(m^{-4})$$

$$\approx 0.693\,147\,18m - 1.039\,720\,77 - 0.002\,697\,58m^{-1} + 0.003\,232\,60m^{-2}$$

$$+ 0.002\,171\,82m^{-3} + O(m^{-4}), \tag{8.11}$$

where $c = \log 2$. In spite of the smallness of the coefficients, there seems to be no general strategy for estimating the remainder after truncating terms in the asymptotic expansion. It is an open question whether the resulting series (in powers of m^{-1}) is purely asymptotic or converges in a neighbourhood of $m = \infty$.

Proof of Theorem 8.7   Since by assumption $N \mid k$, we can write $k = Nk_1$ and thus rewrite (8.5) as

$$0 < \frac{\log 2}{2N} - \frac{k_1}{2m - 3} < \frac{0.0111}{2N(2m - 3)^2}. \tag{8.12}$$

We infer that $k_1/(2m - 3) = p_n/q_n$ is a convergent to $(\log 2)/(2N)$ with n even. Since $p \mid m$ implies that $p - 1 \nmid k$ (see e.g. [113, Proposition 9]), we have $(6, q_n) = 1$. We rewrite (8.12) as

$$0 < \frac{\log 2}{2N} - \frac{p_n}{q_n} < \frac{0.0111}{2Nd^2 q_n^2},$$

with d the greatest common divisor of k_1 and $2m - 3$. However,

$$\frac{\log 2}{2N} - \frac{p_n}{q_n} > \frac{1}{(a_{n+1} + 2)q_n^2}.$$

Hence $(a_{n+1} + 2)^{-1} < 0.0111/(2Nd^2)$, from which the result follows on noting that $2m - 3 \geq q_n$. $\qquad\qquad\square$

8.3 Efficient ways of computing continued fractions

The naïve way to compute the continued fraction for a real number α is to find an accurate numerical approximation to α and then to use the well-known algorithm in Section 2.5, each step of which computes the next partial quotient $a = \lfloor \alpha \rfloor$ and the next complete quotient $\alpha' = 1/(\alpha - a)$. In the course of this process some precision may be lost, and one has to take precautions to stop the algorithm before the partial quotients become incorrect. Lehmer [96] gave a safe stopping criterion and a trick to reduce the amount of multi-length arithmetic; this leads to the so-called *indirect algorithm* [153]. Schönhage [150] (see also [112]) described an algorithm for computing the greatest common divisor of u and v, and the related continued fraction expansion of u/v, in $O(d \log^2 d \, \log \log d)$ steps if neither u nor v exceed 2^d.

A disadvantage of this basic method is that if one wishes to extend the list of partial quotients computed from an initial approximation of α, one has to compute a more accurate initial approximation of α, then compute the new complete quotient using this approximation and the partial quotients already computed from the old approximation and after that extend the list of partial quotients using the new complete quotient. In [18] and [153] a formula is given for computing a rational approximation of the next complete convergent from the first n partial quotients. From that complete convergent some n new partial quotients can be computed. Therefore, each step provides an approximate doubling of the number of partial quotients; this was called by Shiu [153] the *direct algorithm*. To start this algorithm, a few partial quotients are computed with the basic method. In [18] this approach is proposed for algebraic numbers (cf. Section 4.1) whereas Shiu [153] also applies it to more general numbers, such as π and $\log 2$, defined as the zeros of analytical functions for which the logarithmic derivative at some rational point can be computed with arbitrary precision. For each of 13 different such numbers Shiu computed 10 000 partial quotients; their frequency distributions were compared with the distribution which almost all numbers should obey (see Chapter 3), and no significant deviations from the Khintchine–Lévy theory were reported.

The aim of this section is to give details about this direct algorithm. In this we follow the original text [35]. First we discuss error control.

When one computes the partial quotients $a_0, a_1, \dots$ from a numerical approximation $\tilde{\alpha}$ of α, one loses precision. The error can be controlled with the help of the following two lemmas; Lemma 8.12 gives a sufficient condition for $\lfloor \tilde{\alpha} \rfloor = \lfloor \alpha \rfloor$ to be true, while Lemma 8.13 gives upper bounds for the relative error in $\tilde{\alpha}' = 1/(\tilde{\alpha} - \lfloor \tilde{\alpha} \rfloor)$ as a function of $\tilde{\alpha}$, of the relative error in $\tilde{\alpha}$ and of $\tilde{\alpha}'$.

Lemma 8.12 *Let $\tilde{\alpha} > 0$ be a numerical (rational) approximation of $\alpha > 0$ with relative error bounded by δ: $\tilde{\alpha} = \alpha(1 + \epsilon)$ with $|\epsilon| < \delta$. If*

$$(\lfloor \tilde{\alpha} \rfloor + 1)\delta < \tilde{\alpha} - \lfloor \tilde{\alpha} \rfloor < 1 - (\lfloor \tilde{\alpha} \rfloor + 1)\delta \qquad (8.13)$$

then $\lfloor \tilde{\alpha} \rfloor = \lfloor \alpha \rfloor$.

Proof We will show that $\lfloor \tilde{\alpha} \rfloor < \alpha < \lfloor \tilde{\alpha} \rfloor + 1$.

Since $\tilde{\alpha} > 0$, it follows from (8.13) that $\delta < 1/2$. Hence $1 - \delta < 1/(1 + \epsilon)$ and we have

$$\tilde{\alpha}(1 - \delta) < \frac{\tilde{\alpha}}{1 + \epsilon} = \alpha.$$

Furthermore $\tilde{\alpha}\delta < (\lfloor \tilde{\alpha} \rfloor + 1)\delta$, so that, by the first inequality in (8.13), we have $\tilde{\alpha}\delta < \tilde{\alpha} - \lfloor \tilde{\alpha} \rfloor$. Together with the above inequality, this gives $\lfloor \tilde{\alpha} \rfloor < \tilde{\alpha}(1 - \delta) < \alpha$.

On the other hand, $1/(1 + \epsilon) < 1/(1 - \delta)$, so

$$\alpha = \frac{\tilde{\alpha}}{1 + \epsilon} < \frac{\tilde{\alpha}}{1 - \delta}.$$

From the second inequality in (8.13) we get $\tilde{\alpha} < (\lfloor \tilde{\alpha} \rfloor + 1)(1 - \delta)$, which gives us $\tilde{\alpha}/(1 - \delta) < \lfloor \tilde{\alpha} \rfloor + 1$. $\square$

Lemma 8.13 *Suppose that the conditions of Lemma* 8.12 *are satisfied. Let*

$$\alpha' = \frac{1}{\alpha - \lfloor \alpha \rfloor} \quad and \quad \tilde{\alpha}' = \frac{1}{\tilde{\alpha} - \lfloor \tilde{\alpha} \rfloor}.$$

Then an upper bound for the relative error in $\tilde{\alpha}'$ with respect to α' is given by $\tilde{\alpha}\tilde{\alpha}'\delta/(1-\delta)$.

Proof We have, for $\delta < 1/2$,

$$\left| \frac{\tilde{\alpha}' - \alpha'}{\alpha'} \right| = \left| \frac{1/(\tilde{\alpha} - \lfloor \tilde{\alpha} \rfloor) - 1/(\alpha - \lfloor \alpha \rfloor)}{1/(\alpha - \lfloor \alpha \rfloor)} \right| = \left| \frac{\alpha - \tilde{\alpha}}{\tilde{\alpha} - \lfloor \tilde{\alpha} \rfloor} \right|$$

$$= \tilde{\alpha}\tilde{\alpha}' \left| 1 - \frac{\alpha}{\tilde{\alpha}} \right| = \tilde{\alpha}\tilde{\alpha}' \left| \frac{\epsilon}{1 + \epsilon} \right| < \tilde{\alpha}\tilde{\alpha}' \frac{\delta}{1 - \delta}. \qquad \square$$

Exercise 2.27 suggests Lehmer's trick [96] for reducing the amount of multi-precision work. Assuming that we have a very accurate rational approximation u/v of the real number $\alpha > 1$ for very large numbers u and v, we can form suitable lower and upper bounds for u/v by taking the first N ($N = 20$, say) binary digits of u and v: if $u, v < 2^d$, take $u_0 = \lfloor u/2^{d-N} \rfloor$ and $v_0 = \lfloor v/2^{d-N} \rfloor$ (increasing N if necessary to avoid the situation $v_0 = 0$) and choose $a = b_0 = u_0$, $b = b_1 = v_0 + 1$ and $c = c_0 = u_0 + 1$, $d = c_1 = v_0$ in Exercise 2.27. Now we compute the partial quotients $a_0, a_1, \ldots, a_m$ of $c/d = c_0/c_1$ and hence of α, as follows:

$$a_i = \left\lfloor \frac{c_i}{c_{i+1}} \right\rfloor, \quad c_{i+2} = c_i - a_i c_{i+1}, \quad b_{i+2} = b_i - a_i b_{i+1}, \quad i = 1, \ldots, m, \quad (8.14)$$

where m is chosen to be the first index for which either $b_{m+3} < 0$ or $b_{m+3} \geq b_{m+2}$. Note that we do not compute the partial quotients of $a/b = b_0/b_1$, since as long as $0 \leq b_{i+2} < b_{i+1}$ we are sure that a_i is also the correct partial quotient of a/b. After performing the computation (8.14) we update the fraction u/v by acknowledging the computed partial quotients $a_0, a_1, \ldots, a_m$. If $m = 0$, using the partial quotient a_0 we replace u/v by $v/(u - a_0 v)$; in matrix notation, we are setting

$$\binom{u}{v} = \begin{pmatrix} 0 & 1 \\ 1 & -a_0 \end{pmatrix} \binom{u}{v}.$$

In general, using the partial quotients $a_0, a_1, \ldots, a_m$, we are setting

$$\binom{u}{v} = \begin{pmatrix} 0 & 1 \\ 1 & -a_m \end{pmatrix} \begin{pmatrix} 0 & 1 \\ 1 & -a_{m-1} \end{pmatrix} \cdots \begin{pmatrix} 0 & 1 \\ 1 & -a_1 \end{pmatrix} \begin{pmatrix} 0 & 1 \\ 1 & -a_0 \end{pmatrix} \binom{u}{v}.$$

The product of 2×2 matrices is built up first and then used to multiply the vector with entries u and v; this is the only high-precision part of the computation.

In what follows, a number α is given as a zero of an analytical function $f(x)$.

The aim of the direct algorithm is to compute a very good rational approximation to the complete quotient α_{n+1} for α when the partial quotients $a_0, a_1, \ldots, a_n$ are known, and from that approximation to compute n partial quotients of α_{n+1}. This is done as follows. From

$$\alpha = [a_0; a_1, \ldots, a_n, \alpha_{n+1}] = \frac{\alpha_{n+1} p_n + p_{n-1}}{\alpha_{n+1} q_n + q_{n-1}}$$

we find that

$$\alpha_{n+1} = \frac{(-1)^{n+1}}{q_n(p_n - \alpha q_n)} - \frac{q_{n-1}}{q_n}.$$

Now using the mean value theorem and the defining equation $f(\alpha) = 0$, we replace the difference $p_n/q_n - \alpha$ by $f(p_n/q_n)/f'(p_n/q_n)$. (Here, as usual, f' is the derivative of f.) We obtain the approximation

$$\alpha_{n+1} \approx \frac{(-1)^{n+1}}{q_n^2} \frac{f'(p_n/q_n)}{f(p_n/q_n)} - \frac{q_{n-1}}{q_n}.$$

The error in this approximation is approximately

$$\frac{|f''(\alpha)|}{q_n^2 |f'(\alpha)|} \approx \frac{|f''(p_n/q_n)|}{q_n^2 |f'(p_n/q_n)|}.$$

From this rational approximation of α_{n+1}, the partial quotients $a_{n+1}, a_{n+2}, \ldots, a_{n+m}, \ldots$ can be computed as long as $q_{n+m} < \lambda q_n^2$ for some small $\lambda = \lambda(\alpha) > 0$. The direct method for computing N partial quotients of the continued fraction expansion of α now reads as follows.

DIRECT ALGORITHM *Step 1.* Use the basic ('naïve') method to compute a *small* number of partial quotients and the corresponding convergents of α, say up to a_n, p_n, q_n.

Step 2. If $p_n q_{n-1} - p_{n-1} q_n \neq (-1)^{n+1}$ then stop. Compute the next rational approximation α' of α_{n+1} by

$$\alpha' = \frac{(-1)^{n+1}}{q_n^2} \frac{f'(p_n/q_n)}{f(p_n/q_n)} - \frac{q_{n-1}}{q_n}. \tag{8.15}$$

Let $Q = \lambda q_n^2$ for some suitable constant $\lambda = \lambda(\alpha)$. Compute the next partial quotients $a_{n+1}, a_{n+2}, \ldots, a_{n+m}, \ldots$ with the basic method (using Exercise 2.27 and (8.14)) as long as $n + m \leq N$ and $q_{n+m} < Q$.

Step 3. Set $n = n + m$. If $n < N$ then go back to step 2.

The number of partial quotients that can be computed in step 2 is roughly equal to n, so that *after* completion of the step the number of partial quotients computed will roughly have doubled compared with *before* step 2. Since (8.15) is very time consuming, it is desirable to choose n in step 1 such that when step 2 is carried out for the last time the value of n is *slightly larger* than $N/2$. At the beginning of the algorithm the behaviour of step 2 may be rather erratic; one should therefore compute sufficiently many partial quotients of α in step 1 to reach the 'stable' behaviour phase of step 2 (this will involve an approximate doubling of the number of partial quotients). In practice, this works for $n \approx 100$, but that may be due to the size of the first few partial quotients of the continued fraction for α.

8.4 Bounds for solutions

In this section we present some computational details from [65] that were used to achieve the proof of Theorem 8.9.

The computation of $(\log 2)/(2N)$ was done in two steps. First, d digits of $\log 2$ were generated using the γ-cruncher of Yee [167]. With this program, Yee and Chan computed 31 billion decimal digits of $\log 2$ in about 24 hours. Second, a rational approximation to $(\log 2)/(2N)$ was determined with a relative error bounded by 10^{-d}. Then partial quotients of the continued fraction for $(\log 2)/(2N)$ were computed: about $0.97d$ of them can be evaluated (see Exercise 3.9), with safe error control from Section 8.3. The recursion $q_{n+1} = a_{n+1}q_n + q_{n-1}$ for $n \geq 0$, where $q_0 = 1$ and $q_{-1} = 0$, was used to maintain a floating point approximation to the q_n (rounded down) and the residues q_n (mod 6).

The basic method described in the first paragraph of Section 8.3 was first used for $N \leq 2^8 \times 3^4$. It was reasonably fast, reaching the benchmark $m > 10^{10^7}$ in four days with 50×10^6 digits of $\log 2$. The bit-complexity of this algorithm (or of the indirect or direct methods from the previous section) is quadratic, and reaching the $m > 10^{10^{10}}$ milestone would take centuries with the current technology.

A faster version of the program was finally used in [65]; the recursive greatest common divisor method was applied. It was adapted for computing a continued fraction by the use of Exercise 2.27 and Lehmer's trick for error control. With it the program leaped over 10^{10^8} in just about one hour. Finally, the result in Theorem 8.9 was established in no more than 10 hours and 3×10^9 digits of $\log 2$ were obtained:

$$N = 2^8 \times 3^5 \times 5^3, \qquad n(N) = 3\,236\,170\,820, \qquad a_{n+1} = 2\,307\,115\,390,$$
$$q_n \approx 5.427815 \times 10^{1\,667\,658\,416}, \qquad q_n \bmod 6 = 1.$$

Notes

An elementary approach to the Erdős–Moser equation was developed by MacMillan and Sondow in [106, 107]. In [106] they give a simple proof that

$$S_k(p) \bmod p = \begin{cases} -1 & \text{if } p-1 \mid k, \\ 0 & \text{otherwise,} \end{cases}$$

where $S_k(m) = \Sigma_k(m-1) = 1^k + 2^k + \cdots + (m-1)^k$. The idea is to use Pascal's identity from the year 1654.

EXERCISE 8.14 (Pascal's identity)   Show that for integers $m \geq 1$ and $a \geq 2$, we have

$$\sum_{j=0}^{m-1} \binom{m}{j} S_j(a) = a^m - 1.$$

Note that Moser's method can be interpreted as considering (8.1) modulo $m \pm 1$ and modulo $2m \pm 1$; MacMillan and Sondow in [107] go on to investigate it modulo $(m-1)^2$ to gain further arithmetic information.

The asymptotics of $k = k(m)$ in (8.8) or (8.11) can be computed *experimentally* using an elementary approach of Zagier from [170, Section 3]. The method is quite general in nature, and is useful in many situations when finding asymptotics rigorously is hard or impossible. We cannot resist the temptation to give some details of Zagier's method here.

Given a function $f = f(m)$, where m is not necessarily an integer, suppose that we expect asymptotic behaviour of the form

$$f(m) = c_0 + \frac{c_1}{m} + \frac{c_2}{m^2} + \frac{c_3}{m^3} + \cdots \quad \text{as } m \to \infty.$$

Choose some moderately small integer $r > 0$ (say, $r = 10$). Then the rth difference of the function

$$m^r f(m) = c_0 m^r + c_1 m^{r-1} + \cdots + c_{r-1} m + c_r + c_{r+1} m^{-1} + \cdots$$

equals $r! c_0 + O(m^{-r-1})$, which tends very rapidly to $r! c_0$ as m gets large. Once c_0 is computed (numerically), one goes on to compute c_1 by applying the same procedure to $m(f(m) - c_0) = c_1 + c_2/m + \cdots$ and then one continues the process iteratively. In a more general situation we may look for asymptotics of the form

$$f(m) = C^m m^\gamma \left(c_0 + \frac{c_1}{m} + \frac{c_2}{m^2} + \frac{c_3}{m^3} + \cdots \right) \quad \text{as } m \to \infty,$$

where C and γ are unknown as well. Then

$$\frac{f(m+1)}{f(m)} = C + \frac{C\gamma}{m} + \cdots \quad \text{as } m \to \infty,$$

so that one can use the strategy above to determine first C (numerically) and γ.

The method was used to verify numerically the first few coefficients produced in the asymptotic relation (8.11).

EXERCISE 8.15 ([69], [156, sequence A027363]) For $m > 1$, the number v_m of lines in hypersurfaces of degree $2m - 3$ of the projective space $\mathbb{P}^m$ can be computed as the coefficient of x^{m-1} in the polynomial

$$(1 - x) \prod_{j=1}^{2m-3} (2m - 3 - j(1 - x)).$$

Show that v_m is always odd and that

$$v_m = \sqrt{\frac{27}{\pi}}(2m - 3)^{2m-7/2}\left(1 - \frac{9}{8m} - \frac{111}{640m^2} - \frac{9\,999}{25\,600m^3} + \cdots\right) \quad \text{as } m \to \infty.$$

In 1972 Bill Gosper introduced exact algorithms for producing the sum, difference, product and ratio of two continued fractions. As described in http://mathworld.wolfram.com/ContinuedFraction.html:

Gosper has invented an algorithm for performing analytic addition, subtraction, multiplication, and division using continued fractions. It requires keeping track of eight integers which are conceptually arranged at the polyhedron vertices of a cube. Although this algorithm has not appeared in print, similar algorithms have been constructed by Vuillemin (1987) and Liardet and Stambul (1998).

9

Irregular continued fractions

We finish our expedition with a look at irregular continued fractions and pass by some classical gems en route.

9.1 General theory

There exist many generalisations of classical continued fractions. One of the most natural generalisations, which admits many applications not only in number theory but also in analysis, is the following *irregular continued fraction*:

$$a_0 + \cfrac{b_1}{a_1 + \cfrac{b_2}{a_2 + \cfrac{b_3}{a_3 + \cfrac{\ddots}{\ddots + a_{n-1} + \cfrac{b_n}{a_n}}}}},$$

which is written as

$$a_0 + \frac{b_1}{|a_1|} + \frac{b_2}{|a_2|} + \frac{b_3}{|a_3|} + \cdots + \frac{b_n}{|a_n|}.$$

For the regular continued fractions considered earlier, we have $b_n = 1$ for all n. An *infinite irregular continued fraction* can then be written as

$$a_0 + \frac{b_1}{|a_1|} + \frac{b_2}{|a_2|} + \frac{b_3}{|a_3|} + \cdots + \frac{b_n}{|a_n|} + \cdots \tag{9.1}$$

and is formalised as follows. For two given sequences of numbers or indeterminates $\{a_n\}_{n=0}^{\infty}$ and $\{b_n\}_{n=1}^{\infty}$, we define rational functions $S_n(x)$ by the rule

$$S_0(x) = a_0 + x, \qquad S_n(x) = S_{n-1}\left(\frac{b_n}{a_n + x}\right), \quad n = 1, 2, 3, \ldots$$

161

By induction on $n = 1, 2, \ldots$ it is not hard to show that

$$S_n(x) = a_0 + \frac{b_1|}{|a_1} + \frac{b_2|}{|a_2} + \frac{b_3|}{|a_3} + \cdots + \frac{b_n|}{|a_n + x}.$$

Therefore, $r_0 = S_0(0) = a_0$ and

$$r_n = S_n(0) = a_0 + \frac{b_1|}{|a_1} + \frac{b_2|}{|a_2} + \frac{b_3|}{|a_3} + \cdots + \frac{b_n|}{|a_n}, \qquad n = 1, 2, 3, \ldots$$

If we assign numerical values to the sequences $\{a_n\}_{n=0}^{\infty}$ and $\{b_n\}_{n=1}^{\infty}$ (assuming that $b_n \neq 0$ for $n \geq 1$) then we can consider the limit

$$\alpha = \lim_{n \to \infty} r_n.$$

If this limit exists α is said to be the *value* of the irregular continued fraction (9.1); the numbers r_n, where $n = 0, 1, 2, \ldots$, are called the nth *convergents*.

As in the case of regular continued fractions, to every continued fraction (9.1) we assign the sequences of numerators $\{p_n\}_{n=-1}^{\infty}$ and denominators $\{q_n\}_{n=-1}^{\infty}$ of the convergents. They are determined by the linear recurrence equations

$$\begin{aligned} p_{-1} = 1, \quad p_0 = a_0, \quad & p_n = a_n p_{n-1} + b_n p_{n-2}, \quad n = 1, 2, \ldots, \\ q_{-1} = 0, \quad q_0 = 1, \quad & q_n = a_n q_{n-1} + b_n q_{n-2}, \quad n = 1, 2, \ldots \end{aligned} \qquad (9.2)$$

(When $b_n \equiv 1$ we have our familiar regular continued fraction recursions.)

The fact that these sequences indeed provide the numerators and denominators of the corresponding convergents is proved in the following statement.

Theorem 9.1 *If p_n and q_n are the sequences generated by (9.2) for a given continued fraction (9.1) and $S_n(x)$ is the above sequence of rational transformations then*

$$S_n(x) = \frac{p_n + p_{n-1}x}{q_n + q_{n-1}x}, \qquad p_n q_{n-1} - p_{n-1} q_n \neq 0, \qquad n = 0, 1, 2, \ldots$$

In particular,

$$r_n = S_n(0) = \frac{p_n}{q_n}, \qquad n = 0, 1, 2, \ldots$$

EXERCISE 9.2   Prove Theorem 9.1 by induction on $n = 0, 1, 2, \ldots$

Once again, a 2×2 matrix approach is useful. From the recurrence relations in (9.2), we see that

$$\begin{pmatrix} a_0 & 1 \\ 1 & 0 \end{pmatrix} \begin{pmatrix} a_1 & 1 \\ b_1 & 0 \end{pmatrix} \cdots \begin{pmatrix} a_n & 1 \\ b_n & 0 \end{pmatrix} = \begin{pmatrix} p_n & p_{n-1} \\ q_n & q_{n-1} \end{pmatrix}.$$

By taking the determinant of both sides, we obtain immediately

Corollary 9.3 *We have*

$$p_n q_{n-1} - p_{n-1} q_n = (-1)^{n-1} \prod_{k=1}^{n} b_k \qquad for\ n = 0, 1, 2, \ldots$$

Summarising, the sequence of convergents of a continued fraction (9.1) is uniquely determined by the sequences $\{p_n\}_{n=-1}^{\infty}$ and $\{q_n\}_{n=-1}^{\infty}$, which, in turn, are constructed by means of the recurrence relations (9.2). As the following theorem shows, the converse holds as well: the sequences (9.2) define the continued fraction (9.1) in a unique way.

Theorem 9.4 *Let $\{p_n\}_{n=-1}^{\infty}$ and $\{q_n\}_{n=-1}^{\infty}$ be two sequences of numbers such that $q_{-1} = 0$, $p_{-1} = q_0 = 1$ and $p_n q_{n-1} - p_{n-1} q_n \neq 0$ for $n = 0, 1, 2, \ldots$ Then there exists a unique continued fraction (9.1) whose nth numerator is b_n and nth denominator is a_n, for each $n \geq 0$. More precisely,*

$$a_0 = p_0, \quad a_1 = q_1, \quad b_1 = p_1 - p_0 q_1,$$

$$a_n = \frac{p_n q_{n-2} - p_{n-2} q_n}{p_{n-1} q_{n-2} - p_{n-2} q_{n-1}},$$

$$b_n = \frac{p_{n-1} q_n - p_n q_{n-1}}{p_{n-1} q_{n-2} - p_{n-2} q_{n-1}}, \qquad n = 2, 3, \ldots$$

Proof Again, the proof is by induction on $n = 0, 1, 2, \ldots$ $\qquad\qquad\qquad\square$

Theorem 9.1 provides us with a simple algorithm for computing the value of an irregular continued fraction. Namely,

$$r_n = \frac{p_n}{q_n} = p_0 + \sum_{l=1}^{n} \left(\frac{p_l}{q_l} - \frac{p_{l-1}}{q_{l-1}} \right)$$

$$= a_0 + \sum_{l=1}^{n} \frac{(-1)^{l-1} \prod_{k=1}^{l} b_k}{q_l q_{l-1}}, \qquad n = 0, 1, 2, \ldots,$$

implying that

$$a_0 + \frac{b_1|}{|a_1} + \frac{b_2|}{|a_2} + \frac{b_3|}{|a_3} + \cdots + \frac{b_n|}{|a_n} + \cdots = a_0 + \sum_{l=1}^{\infty} \frac{(-1)^{l-1} \prod_{k=1}^{l} b_k}{q_l q_{l-1}}.$$

Therefore, the convergence problem for continued fractions of the form (9.1) can be reduced to a convergence problem for the corresponding series.

Two (irregular) continued fractions

$$a_0 + \frac{b_1|}{|a_1} + \frac{b_2|}{|a_2} + \cdots + \frac{b_n|}{|a_n} + \cdots \quad \text{and} \quad a_0' + \frac{b_1'|}{|a_1'} + \frac{b_2'|}{|a_2'} + \cdots + \frac{b_n'|}{|a_n'} + \cdots$$

$$(9.3)$$

with corresponding sequences of convergents $\{r_n\}_{n=0}^{\infty}$ and $\{r_n'\}_{n=0}^{\infty}$, respectively, are said to be *equivalent* if

$$r_n = r_n' \qquad \text{for all } n = 0, 1, 2, \ldots$$

Theorem 9.5 *Two continued fractions* (9.3) *are equivalent iff there exists a sequence of nonzero numbers* $\{c_n\}_{n=0}^{\infty}$ *with* $c_0 = 1$ *such that*

$$a_n' = c_n a_n, \quad n = 0, 1, 2, \ldots, \qquad b_n' = c_n c_{n-1} b_n, \quad n = 1, 2, \ldots \qquad (9.4)$$

Proof First, assume that relations (9.4) hold. Then it can be easily shown by induction on $n = 0, 1, 2, \ldots$, with the help of the recurrence relations (9.2), that

$$p_n' = p_n \prod_{l=0}^{n} c_l, \qquad q_n' = q_n \prod_{l=0}^{n} c_l, \qquad (9.5)$$

implying that $r_n' = p_n'/q_n' = p_n/q_n = r_n$. Second, if $r_n' = r_n$ for all $n = 0, 1, 2, \ldots$ then take $c_0 = 1$ and define recursively $c_n = p_n'/(p_n \prod_{l=0}^{n-1} c_l)$. Now we arrive at the relations in (9.5), which imply (9.4) in accordance with the formulae of Theorem 9.4. $\qquad\qquad\square$

Finally, we stress that the value of an infinite *irregular* continued fraction is not necessarily an irrational number even when the a_k and b_k are required to be positive integers. An example is given in the following exercise.

EXERCISE 9.6 Compute the value of the continued fraction

$$1 + \frac{2}{\vert 1} + \frac{2}{\vert 1} + \frac{2}{\vert 1} + \cdots + \frac{2}{\vert 1} + \cdots .$$

9.2 Euler continued fraction

Finite identities such as

$$a_0 + a_1 + a_1 a_2 + a_1 a_2 a_3 + a_1 a_2 a_3 a_4 = a_0 + \frac{a_1}{\vert 1} + \frac{a_2}{\vert 1 + a_2} + \frac{a_3}{\vert 1 + a_3} + \frac{a_4}{\vert 1 + a_4}$$

are easily verified symbolically. The general form

$$a_0 + a_1 + a_1 a_2 + a_1 a_2 a_3 + \cdots + a_1 a_2 a_3 \cdots a_N$$

$$= a_0 + \frac{a_1}{\vert 1} + \frac{a_2}{\vert 1 + a_2} + \frac{a_3}{\vert 1 + a_3} + \cdots + \frac{a_N}{\vert 1 + a_N} \qquad (9.6)$$

can then be obtained by substituting $a_N + a_N a_{N+1}$ for a_N and checking that the form of the right-hand side is preserved.

Equation (9.6) allows many series to be re-expressed as irregular continued fractions. For example, with $a_0 = 0$, $a_1 = z$, $a_2 = -z^2/3$, $a_3 = -3z^2/5$, ..., we find that

$$\arctan z = z - \frac{z^3}{3} + \frac{z^5}{5} - \frac{z^7}{7} + \frac{z^9}{9} - \cdots \qquad (9.7)$$

with $|z| \leq 1$, can be expressed as an irregular continued fraction due to Euler:

$$\arctan z = \frac{z}{\vert 1} + \frac{z^2}{\vert 3 - z^2} + \frac{9z^2}{\vert 5 - 3z^2} + \frac{25z^2}{\vert 7 - 5z^2} + \cdots .$$

When $z = 1$, this becomes what is now viewed as the first infinite continued fraction, given by Lord Brouncker (1620–1684):

$$\frac{4}{\pi} = 1 + \frac{1}{\vert 2} + \frac{3^2}{\vert 2} + \frac{5^2}{\vert 2} + \frac{7^2}{\vert 2} + \frac{9^2}{\vert 2} + \cdots . \qquad (9.8)$$

Brouncker intuited this result from Stirling's work on the factorial.

EXERCISE 9.7 (see also [74]) Legitimate the derivation of Brouncker's irregular fraction (9.8) for $4/\pi$.

Furthermore, since

$$\arctan z = \frac{\log(1 + iz) - \log(1 - iz)}{2i},$$

we also obtain a variant of Euler's continued fraction for $\log((1 + z)/(1 - z))$.

EXERCISE 9.8 ([74]) Find the value of the continued fraction

$$\frac{1}{\vert 1} + \frac{1^2}{\vert 1} + \frac{2^2}{\vert 1} + \frac{3^2}{\vert 1} + \frac{4^2}{\vert 1} + \frac{5^2}{\vert 1} + \cdots .$$

While elegant, Euler's continued fraction is much less useful than that of Gauss, to which we now turn.

9.3 Gauss continued fraction for the hypergeometric function

A classical result on an irregular continued fraction for the so-called *hypergeometric function* goes back to Gauss. The function is defined by the series

$$F(a, b; c; z) = \sum_{n=0}^{\infty} \frac{(a)_n (b)_n}{n!(c)_n} z^n$$

$$= 1 + \frac{a \times b}{1 \times c} z + \frac{a(a+1) \times b(b+1)}{1 \times 2 \times c(c+1)} z^2$$

$$+ \frac{a(a+1)(a+2) \times b(b+1)(b+2)}{1 \times 2 \times 3 \times c(c+1)(c+2)} z^3 + \cdots,$$

where the notation

$$(a)_n = \frac{\Gamma(a+n)}{\Gamma(a)} = a(a+1)\cdots(a+n-1)$$

stands for the *Pochhammer symbol* (or *shifted factorial*; note that $(1)_n = n!$). It is not hard to check that the series converges for $|z| < 1$. Among the many properties possessed by the function $F(z) = F(a, b; c; z)$, the fact that it satisfies a second-order linear homogeneous differential equation,

$$z(1-z) \frac{d^2 F}{dz^2} + (c - (a+b+1)z) \frac{dF}{dz} - abF = 0,$$

is crucial. Using this relation one can efficiently construct the analytic continuation of the function $F(z)$, originally defined by the series above, to the whole complex plane with a branch cut along the real ray $[1, \infty)$.

An important feature of the hypergeometric function (and its generalisations) is that it encompasses many other functions, including elementary ones, as either special or limiting cases. For example,

$$\log(1+z) = zF(1, 1; 2; -z), \qquad (1+z)^{-a} = F(a, b; b; -z),$$

$$e^z = \lim_{b \to \infty} F(1, b; 1; z/b), \qquad \arcsin z = zF(1/2, 1/2; 3/2; z^2)$$

$$\text{and} \qquad \arctan z = zF(1, 1/2; 3/2; -z^2).$$

Lemma 9.9 *The following* contiguous *relation holds:*

$$F(a, b; c; z) - F(a+1, b; c+1; z) = \frac{(a-c)b}{c(c+1)} zF(a+1, b+1; c+2; z). \quad (9.9)$$

Proof Indeed, we have

$$F(a, b; c; z) - F(a + 1, b; c + 1; z)$$

$$= \sum_{n=0}^{\infty} \frac{(a)_n (b)_n}{n!(c + 1)_n} \left(\frac{c + n}{c} - \frac{a + n}{a} \right) z^n$$

$$= \sum_{n=1}^{\infty} \frac{(a)_n (b)_n}{n!(c + 1)_n} \frac{(a - c)n}{ac} z^n$$

$$= \sum_{m=0}^{\infty} \frac{(a)_{m+1} (b)_{m+1}}{m!(c + 1)_{m+1}} \frac{a - c}{ac} z^{m+1}$$

$$= \frac{(a - c)bz}{c(c + 1)} \sum_{m=0}^{\infty} \frac{(a + 1)_m (b + 1)_m}{m!(c + 2)_m} z^m. \qquad \square$$

Note that, because of the symmetry between a and b in the hypergeometric function $F(a, b; c; z)$, we also obtain from (9.9) another contiguous relation:

$$F(a, b; c; z) - F(a, b + 1; c + 1; z) = \frac{a(b - c)}{c(c + 1)} zF(a + 1, b + 1; c + 2; z). \quad (9.10)$$

Theorem 9.10 (Gauss continued fraction) *We have*

$$\frac{F(a + 1, b; c + 1; z)}{cF(a, b; c; z)} = \frac{1}{\lvert c} + \frac{\lambda_0 z}{\lvert c + 1} + \frac{\lambda_1 z}{\lvert c + 2} + \cdots + \frac{\lambda_n z}{\lvert c + n + 1} + \cdots, \quad (9.11)$$

where $\lambda_{2k-1} = (a + k)(b - c - k)$ *and* $\lambda_{2k} = (a - c - k)(b + k)$.

Proof For $k = 0, 1, 2, \ldots$, define

$$F_{2k}(z) = F(a + k, b + k; c + 2k; z)$$

and

$$F_{2k+1}(z) = F(a + k + 1, b + k; c + 2k + 1; z).$$

Then

$$F_{2k}(z) - F_{2k+1}(z) = \frac{(a - c - k)(b + k)}{(c + 2k)(c + 2k + 1)} zF_{2k+2}(z),$$

$$F_{2k-1}(z) - F_{2k}(z) = \frac{(a + k)(b - c - k)}{(c + 2k - 1)(c + 2k)} zF_{2k+1}(z),$$

by the contiguous relations (9.9) and (9.10), respectively. Therefore

$$\frac{F_{n+1}(z)}{F_n(z)} = \frac{1}{1 + \dfrac{\lambda_n z}{(c + n)(c + n + 1)} \dfrac{F_{n+2}(z)}{F_{n+1}(z)}},$$

where $\lambda_{2k} = (a - c - k)(b + k)$ and $\lambda_{2k-1} = (a + k)(b - c - k)$, so that

$$\frac{F_1(z)}{F_0(z)} = \frac{1}{|1} + \frac{\left|\frac{\lambda_0 z}{c(c+1)}\right.}{|1} + \frac{\left|\frac{\lambda_1 z}{(c+1)(c+2)}\right.}{|1} + \cdots + \frac{\left|\frac{\lambda_n z}{(c+n)(c+n+1)}\right.}{|1} + \cdots .$$

It remains to pass to the equivalent continued fraction by taking $c_n = c + n - 1$ for $n = 1, 2, \ldots$ in the notation of Theorem 9.5.   □

Our derivation of the Gauss continued fraction follows the lines of Section 2.10, where we derived a continued fraction for a confluent hypergeometric function, also known as Bessel's function. By taking the limit $a \to 0$ in (9.11) and specialising to $c = b$, we obtain the continued fraction

$$\sum_{n=0}^{\infty} \frac{z^n}{b + n} = \frac{F(1, b; b + 1; z)}{b}$$

$$= \frac{1}{|b} + \frac{\lambda_0' z}{|b + 1} + \frac{\lambda_1' z}{|b + 2} + \cdots + \frac{\lambda_n' z}{|b + n + 1} + \cdots , \qquad (9.12)$$

where $\lambda_{2k-1}' = -k^2$ and $\lambda_{2k}' = -(b + k)^2$; this can be used to construct continued fractions for

$$\log(1 + z) = zF(1, 1; 2; -z) \qquad \text{and} \qquad \arctan z = zF(1, 1/2; 3/2; -z^2),$$

as well as for the tails of their power series.

Theorem 9.11   *For $N = 1, 2, \ldots$, we have*

$$\log(1 + z) - \sum_{n=1}^{N-1} \frac{(-1)^{n-1} z^n}{n} = \frac{(-1)^{N-1} z^N}{|N} + \frac{N^2 z}{|N + 1} + \frac{1^2 z}{|N + 2}$$

$$+ \frac{(N + 1)^2 z}{|N + 3} + \frac{2^2 z}{|N + 4} + \cdots , \qquad (9.13)$$

$$\arctan z - \sum_{n=0}^{N-1} \frac{(-1)^n z^{2n+1}}{2n + 1} = \frac{(-1)^N z^{2N+1}}{|2N + 1} + \frac{(2N + 1)^2 z^2}{|2N + 3} + \frac{2^2 z^2}{|2N + 5}$$

$$+ \frac{(2N + 3)^2 z^2}{|2N + 7} + \frac{4^2 z^2}{|2N + 9} + \cdots . \qquad (9.14)$$

Proof   Use

$$\log(1 + z) - \sum_{n=1}^{N-1} \frac{(-1)^{n-1} z^n}{n} = \frac{z^N F(1, N; N + 1; -z)}{N}$$

and

$$\arctan z - \sum_{n=0}^{N-1} \frac{(-1)^n z^{2n+1}}{2n + 1} = \frac{z^{2N+1} F(1, N + 1/2; N + 2/2; -z^2)}{2(N + 1/2)}$$

together with the continued fraction (9.12).   □

Note that further specialisation to $z = 1$ in (9.14) leads to a continued fraction for the tail of the approximation of $\pi/4$ by Gregory's series:

$$
\frac{\pi}{4} - \sum_{n=0}^{N-1} \frac{(-1)^n}{2n+1} = \frac{(-1)^N}{\big\lceil 2N+1} + \frac{(2N+1)^2}{\big\lceil 2N+3} + \frac{2^2}{\big\lceil 2N+5}
$$

$$
+ \frac{(2N+3)^2}{\big\lceil 2N+7} + \frac{4^2}{\big\lceil 2N+9} + \cdots , \qquad (9.15)
$$

while its special case $N = 0$ gives

$$
\pi = \frac{4}{\big\lceil 1} + \frac{1^2}{\big\lceil 3} + \frac{2^2}{\big\lceil 5} + \frac{3^2}{\big\lceil 7} + \cdots + \frac{n^2}{\big\lceil 2n+1} + \cdots . \qquad (9.16)
$$

The estimates given in [25, Theorem 4] show that, for large b and z from the range $-1 \le z < 0$, the quality of approximation of the hypergeometric value by the nth convergent from (9.12) is roughly bounded by $(z/(z-1))^n$. The estimate is applicable in the case of (9.15), thus showing that the convergents to the continued fraction can be used to accelerate the convergence of Gregory's series. Note that the paper [25] discusses an interesting phenomenon concerning the tail (9.15) of Gregory's series, relating to its asymptotic power series in $1/N$ as $N \to \infty$.

Finally, we mention that, on using the Nth convergent of the continued fraction in (9.13) for $z = 1$, it is possible to demonstrate that the resulting rational approximations to $\log 2$ are sufficient for proving the irrationality of the number (using Theorem 1.34) as well as for producing a bound for its irrationality exponent (using Theorem 1.35). We will not pursue this topic further here, but the idea is developed by M. Prévost in [137] for irrationality proofs of the constants $\zeta(2)$ and $\zeta(3)$.

9.4 Ramanujan's AGM continued fraction

In this section we give a taste of Ramanujan's AGM continued fraction; the more famous Rogers–Ramanujan continued fraction is discussed in the chapter notes. The substantial technical details can be found in the papers [26, 27, 19] and [102].

In [26, 27] one of the present authors considered the arithmetic-geometric mean (AGM) fraction, found in Chapter 18 of Ramanujan's second notebook

[13]. For $a, b, \eta > 0$ we have

$$\mathcal{R}_\eta(a, b) = \cfrac{a}{\eta + \cfrac{b^2}{\eta + \cfrac{4a^2}{\eta + \cfrac{9b^2}{\eta + \ddots}}}}, \qquad (9.17)$$

one of whose remarkable properties is a formal AGM relation that is known to be true at least for positive real a, b:

$$\mathcal{R}_1\left(\frac{a+b}{2}, \sqrt{ab}\right) = \frac{\mathcal{R}_1(a, b) + \mathcal{R}_1(b, a)}{2}. \qquad (9.18)$$

However, this relation is of dubious validity for general complex parameters [26] despite claims in the literature. Note that $\mathcal{R}_\eta(a, b) = \mathcal{R}_1(a/\eta, b/\eta)$ and that the validity of (9.18) depends only on the ratio a/b, which can be thought as a point of the extended complex field $\overline{\mathbb{C}} = \mathbb{C} \cup \{\infty\}$.

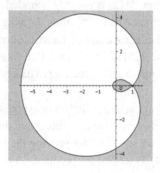

Figure 9.1 Cardioid on whose complement (shaded) AGM relation (9.18) holds.

The work in [27] focused on the convergence domain

$$\mathcal{D}_0 = \{(a, b) \in \mathbb{C}^2 : \mathcal{R}_1(a, b) \text{ converges on } \overline{\mathbb{C}}\}.$$

It was proved therein that, with

$$\mathcal{D}_1 = \{(a, b) \in \mathbb{C}^2 : |a| \neq |b|\} \cup \{(a, b) \in \mathbb{C}^2 : a^2 = b^2 \notin (-\infty, 0)\},$$

we have $\mathcal{D}_1 \subseteq \mathcal{D}_0$, so that the Ramanujan continued fraction converges for almost all complex pairs (a, b). The article [19] showed that $\mathcal{D}_1 = \mathcal{D}_0$. Equivalently, the fraction $\mathcal{R}_1$ diverges whenever $0 \neq a = be^{i\phi}$ with $\cos^2 \phi \neq 1$ or $a^2 = b^2 \in (-\infty, 0)$. It was also shown that remarkable and explicit chaotic dynamics occur on the imaginary axis, say for $\mathcal{R}_1(i, i)$. Key to all this analysis is the following theorem, whose components date back to the nineteenth century. The notation

$$\vartheta_2(q) = \sum_{\substack{k \in \mathbb{Z} \\ k \text{ odd}}} q^{k^2/4} = 2q^{1/4} \sum_{n=0}^{\infty} q^{n(n+1)}, \qquad \vartheta_3(q) = \sum_{\substack{k \in \mathbb{Z} \\ k \text{ even}}} q^{k^2/4} = 1 + 2 \sum_{n=1}^{\infty} q^{n^2}$$

stands for the classical theta (null) functions [166].

Theorem 9.12 ([26, 19]) *For real $y, \eta > 0$ and $q = e^{-\pi y}$, we have the theta function parametrisations*

$$\eta \sum_{\substack{k \in \mathbb{Z} \\ k \text{ odd}}} \frac{\operatorname{sech}(k\pi y/2)}{\eta^2 + k^2} = \mathcal{R}_\eta(\vartheta_2^2(q), \vartheta_3^2(q)),$$

$$\eta \sum_{\substack{k \in \mathbb{Z} \\ k \text{ even}}} \frac{\operatorname{sech}(k\pi y/2)}{\eta^2 + k^2} = \mathcal{R}_\eta(\vartheta_3^2(q), \vartheta_2^2(q)).$$

Moreover, the equality (9.18) *holds when a/b belongs to the closed exterior of the cardioid knot shown in Figure* 9.1, *which in polar coordinates is given by $r^2 + (2 \cos \phi - 4)r + 1 = 0$.*

Interpreting Theorem 9.12 as giving a Riemann integral in the limit $b \to a^-$ (for $a > 0$), gives a slew of relations involving the *psi* or *digamma* function

$$\psi(z) = \frac{\mathrm{d}}{\mathrm{d}z} \log \Gamma(z),$$

the hypergeometric function and the Gauss continued fraction (see Section 9.3).

Corollary 9.13 ([26]) *For all $a > 0$,*

$$\mathcal{R}_1(a, a) = \int_0^{\infty} \frac{\operatorname{sech}(\pi x/(2a))}{1 + x^2} \, \mathrm{d}x$$

$$= 2a \sum_{k=1}^{\infty} \frac{(-1)^{k+1}}{1 + (2k - 1)a}$$

$$= \frac{1}{2} \left(\psi\left(\frac{3}{4} + \frac{1}{4a}\right) - \psi\left(\frac{1}{4} + \frac{1}{4a}\right) \right)$$

$$= \frac{2a}{1 + a} F\left(\frac{1}{2a} + \frac{1}{2}, 1; \frac{1}{2a} + \frac{3}{2}; -1\right)$$

$$= 2 \int_0^1 \frac{t^{1/a}}{1 + t^2} \, \mathrm{d}t$$

$$= \int_0^{\infty} e^{-x/a} \operatorname{sech} x \, \mathrm{d}x.$$

No closed form is known for any case with $a \neq b$. In [29, 30] various extensions were studied and fractions with period 3 and other features were elaborated.

EXERCISE 9.14   Derive a restricted parametrised form of (9.18) from Theorem 9.12 (as $q \mapsto q^2$).

EXERCISE 9.15   Derive Corollary 9.13 from Theorem 9.12.

Hint   The expressions are listed in an order suited to proving each expression from the previous one.   □

EXERCISE 9.16 ([26])   Use Corollary 9.13 to determine that $\mathcal{R}_1(1,1) = \log 2$ and $\mathcal{R}_1(1/2, 1/2) = 2 - \pi/2$. Obtain for positive integers p, q that

$$\mathcal{R}_1\!\left(\frac{p}{q}, \frac{p}{q}\right) = -2p \sum_{n=1}^{p+q-1} \frac{1}{n}\big(\delta_{n \equiv p+q \ (\mathrm{mod}\, 4p)} - \delta_{n \equiv 3p+q \ (\mathrm{mod}\, 4p)}\big)$$

$$- 2 \sum_{\substack{0<k<2p \\ k \ \mathrm{odd}}} \cos\frac{(p+q)k\pi}{2p} \, \log\!\left(2\sin\frac{k\pi}{4p}\right)$$

$$+ 2\pi \sum_{\substack{0<k<2p \\ k \ \mathrm{odd}}} \left(\frac{1}{2} - \frac{k}{4p}\right) \sin\frac{(p+q)k\pi}{2p},$$

where δ_X denotes the indicator of set X.

The convergence of (9.17) is slowest – at an arithmetic rate – when $a = b$. The key to analysing the AGM fraction is the replacement of the irregular fraction by a *reduced fraction*, in which, as observed earlier, we have the same form as in a regular fraction except that we require that a_n be real or complex.

EXERCISE 9.17   Show that for all a, b we have

$$\mathcal{R}_1(a, b) = \frac{a}{[a_1, a_2, \ldots, a_n, \ldots]},$$

where

$$a_n = \frac{n!^2}{(n/2)!^4}\, 4^{-n}\, \frac{b^n}{a^n} \approx \frac{2}{\pi n}\, \frac{b^n}{a^n} \qquad \text{for even } n,$$

$$a_n = \frac{((n-1)/2)!^4}{n!^2}\, 4^{n-1}\, \frac{a^{n-1}}{b^{n+1}} \approx \frac{\pi}{2abn}\, \frac{a^n}{b^n} \qquad \text{for odd } n,$$

and so the series $\sum_{n=1}^{\infty} a_n$ diverges.

Now, the Seidel–Stern theorem [82, 103] asserts that for reduced fractions with positive terms the sum diverges iff the fraction converges. This shows that for real $a, b > 0$ the continued fraction $\mathcal{R}_1(a, b)$ converges.

Effective algorithms for computing (9.17) in the full complex plane are given in [26, 27]. They produce D good digits for $O(D)$ operations where the order constant D is independent of a, b, η.

9.5 An irregular continued fraction for $\zeta(2) = \pi^2/6$

Recall the definition of the Riemann zeta function from (1.8). In this section we will construct an irregular continued fraction for the number

$$\zeta(2) = \sum_{n=1}^{\infty} \frac{1}{n^2}.$$

The fact that $\zeta(2) = \pi^2/6$ is known from analysis. For example, it follows from the Fourier expansion of the function x^2 or from the product formula for the function $\sin x$; see [157, 28].

EXERCISE 9.18 ([28])   Show by elementary methods that

$$\left(\sum_{n=-\infty}^{\infty} \frac{(-1)^n}{2n+1} \right)^2 = \sum_{n=-\infty}^{\infty} \frac{1}{(2n+1)^2}.$$

Deduce from this that $\zeta(2) = \pi^2/6$.

Hint   Let

$$\delta_N = \sum_{m,n=-N}^{N} \frac{(-1)^{n+m}}{(2n+1)(2m+1)} - \sum_{k=-N}^{N} \frac{1}{(2k+1)^2} \quad \text{and} \quad \varepsilon_N = \sum_{\substack{m,n=-N \\ m \neq n}}^{N} \frac{1}{m-n}.$$

Show that $\varepsilon_N \leq 1/(N-n+1)$, while

$$\delta_N = \sum_{\substack{m,n=-N \\ m \neq n}}^{N} \frac{(-1)^{n+m}}{(2n+1)(m-n)} \to 0. \qquad \square$$

For each $n = 0, 1, 2, \ldots$, define the rational function

$$R_n(t) = (-1)^n \frac{n! \prod_{j=1}^{n}(t-j)}{\prod_{j=0}^{n}(t+j)^2}$$

and consider the quantity

$$r_n = \sum_{v=1}^{\infty} R_n(v).$$

The latter series converges (absolutely), since $R_n(t) = O(t^{-2})$ as $t \to \infty$.

Lemma 9.19 *The following representation holds:*

$$r_n = q_n \zeta(2) - p_n, \qquad n = 0, 1, 2, \ldots,$$

where

$$q_n = \sum_{k=0}^{n} (R_n(t)(t+k)^2)\big|_{t=-k} \in \mathbb{Q} \qquad and \qquad p_n \in \mathbb{Q}, \qquad n = 0, 1, 2, \ldots$$

In addition, for $n = 0$ and $n = 1$ we have $r_0 = \zeta(2)$ and $r_1 = 3\zeta(2) - 5$, that is,

$$p_0 = 0, \quad q_0 = 1 \qquad and \qquad p_1 = 5, \quad q_1 = 3.$$

Proof First, consider the particular cases $n = 0$ and $n = 1$. For $n = 0$ we have $R_0(t) = 1/t^2$, and hence

$$r_0 = \sum_{v=1}^{\infty} R_0(v) = \zeta(2).$$

For $n = 1$ we decompose the function $R_1(t)$ into a sum of partial fractions:

$$R_1(t) = -\frac{t-1}{t^2(t+1)^2} = \frac{1}{t^2} + \frac{2}{(t+1)^2} - \frac{3}{t} + \frac{3}{t+1}.$$

Thus,

$$r_1 = \sum_{v=1}^{\infty} R_1(v) = \sum_{v=1}^{\infty}\left(\frac{1}{t^2} + \frac{2}{(t+1)^2} - \frac{3}{t} + \frac{3}{t+1}\right)$$

$$= \sum_{v=1}^{\infty} \frac{1}{v^2} + \sum_{v=2}^{\infty} \frac{2}{v^2} + \sum_{v=1}^{\infty}\left(\frac{3}{v+1} - \frac{3}{v}\right)$$

$$= \zeta(2) + 2(\zeta(2) - 1) - 3 = 3\zeta(2) - 5.$$

In the general case, let us replace the quantity r_n with the power series

$$r_n(z) = \sum_{v=1}^{\infty} R_n(v)\, z^v,$$

which converges at $z = 1$ by the argument indicated above. The partial-fraction decomposition of $R_n(t)$ is as follows:

$$R_n(t) = \sum_{k=0}^{n}\left(\frac{A_k}{(t+k)^2} + \frac{B_k}{t+k}\right),$$

where A_k and B_k are certain rational numbers. For the moment, we need 'explicit' formulae for the coefficients

$$A_k = (R_n(t)(t+k)^2)\big|_{t=-k}, \qquad k = 0, 1, \ldots, n,$$

only. We obtain

$$
\begin{aligned}
r_n(z) &= \sum_{\nu=1}^{\infty} R_n(\nu) z^{\nu} = \sum_{\nu=1}^{\infty} \sum_{k=0}^{n} \frac{A_k z^{\nu}}{(\nu+k)^2} + \sum_{\nu=1}^{\infty} \sum_{k=0}^{n} \frac{B_k z^{\nu}}{\nu+k} \\
&= \sum_{k=0}^{n} A_k z^{-k} \sum_{\nu=1}^{\infty} \frac{z^{\nu+k}}{(\nu+k)^2} + \sum_{k=0}^{n} B_k z^{-k} \sum_{\nu=1}^{\infty} \frac{z^{\nu+k}}{\nu+k} \\
&= \sum_{k=0}^{n} A_k z^{-k} \left(\sum_{l=1}^{\infty} \frac{z^l}{l^2} - \sum_{l=1}^{k} \frac{z^l}{l^2} \right) + \sum_{k=0}^{n} B_k z^{-k} \left(\sum_{l=1}^{\infty} \frac{z^l}{l} - \sum_{l=1}^{k} \frac{z^l}{l} \right) \\
&= A(z) \sum_{l=1}^{\infty} \frac{z^l}{l^2} + B(z) \sum_{l=1}^{\infty} \frac{z^l}{l} - C(z) \\
&= A(z) \sum_{l=1}^{\infty} \frac{z^l}{l^2} + B(z)(-\log(1-z)) - C(z),
\end{aligned}
$$

where

$$
A(z) = \sum_{k=0}^{n} A_k z^{-k} \in \mathbb{Q}[z^{-1}], \qquad B(z) = \sum_{k=0}^{n} B_k z^{-k} \in \mathbb{Q}[z^{-1}],
$$

$$
C(z) = \sum_{k=0}^{n} A_k z^{-k} \sum_{l=1}^{k} \frac{z^l}{l^2} + \sum_{k=0}^{n} B_k z^{-k} \sum_{l=1}^{k} \frac{z^l}{l} \in \mathbb{Q}[z^{-1}].
$$

Since the power series $r_n(z)$ converges at $z = 1$, we can use Abel's theorem, which says that the right-hand side of the resulting series $r_n(z)$ has a finite limit r_n as $z \to 1$. In particular, this means that $B(1) = 0$ and $r_n = r_n(1) = A(1)\zeta(2) - C(1)$. It remains to take $q_n = A(1) = \sum_{k=0}^{n} A_k$ and $p_n = C(1)$. $\quad\square$

Note that the function $R_{n+1}(t)/R_n(t)$ is a rational function not only of the parameter t but also of n. Let us define another (rational) function $S_n(t) = s_n(t)R_n(t)$, where

$$
s_n(t) = 11n^2 + 9n + 2 + 3(2n+1)t - t^2. \tag{9.19}
$$

Lemma 9.20 *For each integer $n = 1, 2, \ldots$, the following identity holds:*

$$
(n+1)^2 R_{n+1}(t) - (11n^2 + 11n + 3)R_n(t) - n^2 R_{n-1}(t) = S_n(t+1) - S_n(t). \tag{9.20}
$$

Proof Since

$$
\frac{R_{n-1}(t)}{R_n(t)} = -\frac{(t+n)^2}{n(t-n)}, \qquad \frac{R_{n+1}(t)}{R_n(t)} = -\frac{(n+1)(t-n-1)}{(t+n+1)^2},
$$

$$
\frac{S_n(t+1)}{R_n(t)} = \frac{S_n(t+1)}{R_n(t+1)} \frac{R_n(t+1)}{R_n(t)} = s_n(t+1)\frac{t^3}{(t-n)(t+n+1)^2},
$$

the proof is reduced to verification of the identity

$$(n+1)^2\left(-\frac{(n+1)(t-n-1)}{(t+n+1)^2}\right) - (11n^2+11n+3) - n^2\left(-\frac{(t+n)^2}{n(t-n)}\right)$$

$$= s_n(t+1)\frac{t^3}{(t-n)(t+n+1)^2} - s_n(t), \qquad (9.21)$$

where the polynomial $s_n(t)$ is given in (9.19). Calculation shows that both sides of (9.21) are equal to

$$\frac{\xi_n(t)}{(t-n)(t+n+1)^2}$$

where

$$\xi_n(t) = nt^4 - (7n^2+9n+3)t^3 - (6n^3+30n^2+27n+7)t^2$$
$$+ (17n^4+24n^3+3n^2-6n-2)t + (11n^4+31n^3+31n^2+13n+2)n. \quad \square$$

Theorem 9.21 *The sequences $\{r_n\}_{n=0}^{\infty}$, $\{q_n\}_{n=0}^{\infty}$ and $\{p_n\}_{n=0}^{\infty}$ each satisfy the recurrence relation*

$$(n+1)^2 r_{n+1} - (11n^2+11n+3)r_n - n^2 r_{n-1} = 0, \qquad n = 1, 2, \dots \quad (9.22)$$

Proof We use the definition of r_n and Lemma 9.20: thus,

$$(n+1)^2 r_{n+1} - (11n^2+11n+3)r_n - n^2 r_{n-1}$$

$$= \sum_{v=1}^{\infty}(S_n(t+1) - S_n(t))\Big|_{t=v} = -S_n(1) = -s_n(1)R_n(1) = 0,$$

because $R_n(1) = 0$ for $n = 1, 2, \dots$

For the sequence of coefficients q_n, we use the formula from the proof of Lemma 9.19:

$$q_n = \sum_{k=0}^{n}(R_n(t)(t+k)^2)\Big|_{t=-k} = \sum_{k\in\mathbb{Z}}(R_n(t)(t+k)^2)\Big|_{t=-k},$$

where we have $(R_n(t)(t+k)^2)\Big|_{t=-k} = 0$ for $k < 0$ and $k > n$ for trivial reasons. With the help of identity (9.20) we find that

$$(n+1)^2 q_{n+1} - (11n^2+11n+3)q_n - n^2 q_{n-1}$$

$$= \sum_{k\in\mathbb{Z}}((S_n(t+1) - S_n(t))(t+k)^2)\Big|_{t=-k}$$

$$= \sum_{k\in\mathbb{Z}}((S_n(l-k+1) - S_n(l-k))l^2)\Big|_{l=0}$$

$$= \left(l^2\sum_{k\in\mathbb{Z}}(S_n(l-k+1) - S_n(l-k))\right)\Big|_{l=0} = 0$$

for $n = 0, 1, 2, \ldots$, since the inner sum over k telescopes.

Finally, the sequence $p_n = q_n \zeta(2) - r_n$ satisfies the required recurrence as a linear combination (with constant coefficients) of the sequences q_n and r_n. $\quad\square$

Lemma 9.22   *For all $n = 0, 1, 2, \ldots$ we have $q_n \geq 1$. Moreover, $r_n \to 0$ as $n \to \infty$.*

Proof   From Lemma 9.19 we have $q_0 = 1$ and $q_1 = 3 > 1$. Hence an inductive argument gives us

$$q_{n+1} = \frac{(11n^2 + 11n + 3)q_n + n^2 q_{n-1}}{(n+1)^2} \geq \frac{(11n^2 + 11n + 3) + n^2}{(n+1)^2} > 1$$

for $n = 1, 2, \ldots$. The fact that $r_n \to 0$ as $n \to \infty$ (indeed, a stronger statement about the asymptotics of the quantities r_n) will be proved in the next section.
$$\qquad\square$$

As a consequence, we derive that

$$\frac{p_n}{q_n} = \zeta(2) - \frac{r_n}{q_n} \to \zeta(2) \qquad \text{as } n \to \infty. \tag{9.23}$$

Thus, we have constructed a sequence of *rational* numbers $\{p_n\}_{n=0}^{\infty}$ and $\{q_n\}_{n=0}^{\infty}$ that satisfy the recurrence relation

$$p_n = \frac{P(n-1)}{n^2} p_{n-1} + \frac{(n-1)^2}{n^2} p_{n-2},$$

$$q_n = \frac{P(n-1)}{n^2} q_{n-1} + \frac{(n-1)^2}{n^2} q_{n-2},$$

where $P(n) = 11n^2 + 11n + 3$, for $n = 2, 3, \ldots$. Setting $p_{-1} = 1$, $q_{-1} = 0$ and taking into account that $p_0 = 0$, $q_0 = 1$ and $p_1 = 5$, $q_1 = 3$, by Lemma 9.19, we conclude that

$$p_1 = 3p_0 + 5p_{-1}, \qquad q_1 = 3q_0 + 5q_{-1}.$$

We are now in a position to apply Theorem 9.1. The continued fraction given by

$$\frac{b_1|}{|a_1} + \frac{b_2|}{|a_2} + \frac{b_3|}{|a_3} + \cdots + \frac{b_n|}{|a_n} + \cdots ,$$

$$\text{with} \qquad a_n = \frac{P(n-1)}{n^2} \qquad \text{for } n = 1, 2, \ldots, \tag{9.24}$$

$$b_1 = 5, \quad b_n = \frac{(n-1)^2}{n^2} \qquad \text{for } n = 2, 3, \ldots,$$

has $\{p_n\}_{n=-1}^{\infty}$ and $\{q_n\}_{n=-1}^{\infty}$ as the sequences of numerators and denominators of its convergents; moreover, we have $p_n/q_n \to \zeta(2)$ as $n \to \infty$ by (9.23). Hence

$\zeta(2)$ is the value of the continued fraction (9.24). Developing the equivalent transformation of the resulted continued fraction (see Theorem 9.5) with the choices $c_0 = 1$ and $c_n = n^2$ for $n = 1, 2, \ldots$, we finally arrive at the continued fraction

$$\zeta(2) = \frac{b_1'|}{|a_1'} + \frac{b_2'|}{|a_2'} + \frac{b_3'|}{|a_3'} + \cdots + \frac{b_n'|}{|a_n'} + \cdots$$

with $\quad a_n' = P(n-1) \quad$ for $n = 1, 2, \ldots,$

$\quad b_1' = 5, \quad b_n' = (n-1)^4 \quad$ for $n = 2, 3, \ldots$

Let us summarise our findings in the following statement.

Theorem 9.23   *We have the following (irregular) continued fraction:*

$$\zeta(2) = \frac{5|}{|3} + \frac{1^4|}{|P(1)} + \frac{2^4|}{|P(2)} + \cdots + \frac{n^4|}{|P(n)} + \cdots,$$

where $P(n) = 11n^2 + 11n + 3$.

EXERCISE 9.24 (Irregular continued fraction for $\zeta(3)$)   Take the rational function

$$\tilde{R}_n(t) = \frac{\prod_{j=1}^n (t-j)^2}{\prod_{j=0}^n (t+j)^2}$$

and, for each $n = 0, 1, 2, \ldots$, consider the (absolutely convergent) series

$$\tilde{r}_n = -\sum_{\nu=1}^\infty \frac{d\tilde{R}_n(t)}{dt}\bigg|_{t=\nu}.$$

(a) Show that

$$\tilde{r}_0 = 2\zeta(3) \quad \text{and} \quad \tilde{r}_1 = 10\zeta(3) - 12.$$

(b) Show that, for each $n = 0, 1, 2, \ldots$, we have $\tilde{r}_n = \tilde{q}_n\zeta(3) - \tilde{p}_n$, where $\tilde{p}_n$ and $\tilde{q}_n$ are rational numbers, $\tilde{q}_n > 0$.

(c) Define $\tilde{S}_n(t) = \tilde{s}_n(t)\tilde{R}_n(t)$, where

$$\tilde{s}_n(t) = 4(2n+1)(-2t^2 + t + (2n+1)^2).$$

Check that

$$(n+1)^3 \tilde{R}_{n+1}(t) - (2n+1)(17n^2 + 17n + 5)\tilde{R}_n(t) + n^3 \tilde{R}_{n-1}(t)$$
$$= \tilde{S}_n(t+1) - \tilde{S}_n(t) \tag{9.25}$$

for $n = 1, 2, \ldots$

(d) Using (c), show that the sequences $\{\tilde{r}_n\}_{n=0}^{\infty}$, $\{\tilde{q}_n\}_{n=0}^{\infty}$ and $\{\tilde{p}_n\}_{n=0}^{\infty}$ each satisfy the recurrence relation

$$(n+1)^3\, \tilde{r}_{n+1} - (2n+1)(17n^2 + 17n + 5)\, \tilde{r}_n + n^3\, \tilde{r}_{n-1} = 0, \qquad n = 1, 2, \ldots$$
$$(9.26)$$

(e) Assuming that $\tilde{r}_n \to 0$ as $n \to \infty$, prove the following continued fraction expansion for $\zeta(3)$:

$$\zeta(3) = \frac{6}{\lfloor 5} + \frac{-1^6}{\lfloor Q(1)} + \frac{-2^6}{\lfloor Q(2)} + \cdots + \frac{-n^6}{\lfloor Q(n)} + \cdots ,$$

where $Q(n) = (2n+1)(17n^2 + 17n + 5)$.

9.6 The irrationality of π^2

The aim of this final section is to prove that $\zeta(2)$ is irrational.

Theorem 9.25 *The number $\zeta(2) = \pi^2/6$ is irrational.*

The proof, which we present below, is based on the original construction of Apéry (who also proved the irrationality of $\zeta(3)$; see Exercises 9.24 and 9.30). However, our ideas differ considerably from those of Apéry. Note that the irrationality problem of the numbers $\zeta(5), \zeta(7), \zeta(9), \ldots$ is not yet resolved.

As in the previous section, to each $n = 0, 1, 2, \ldots$ we assign the rational function

$$R_n(t) = (-1)^n \frac{n!\, \prod_{j=1}^{n}(t - j)}{\prod_{j=0}^{n}(t + j)^2}$$

and the corresponding quantity

$$r_n = \sum_{v=1}^{\infty} R_n(v) = q_n \zeta(2) - p_n.$$

Let $d_n = \mathrm{lcm}(1, 2, \ldots, n)$. The corollary of the prime number theorem (see Theorem 1.20) asserts that

$$\lim_{n\to\infty} \frac{\log d_n}{n} = 1, \qquad (9.27)$$

in other words, that d_n grows with n as $e^{n+o(n)}$.

Lemma 9.26 *The rational coefficients in the partial-fraction decomposition*

$$R_n(t) = \sum_{k=0}^{n} \left(\frac{A_k}{(t + k)^2} + \frac{B_k}{t + k} \right)$$

satisfy the inclusions $A_k \in \mathbb{Z}$ and $d_n B_k \in \mathbb{Z}$ for $k = 0, 1, \ldots, n$.

Proof   Write $R_n(t)$ as a product of two 'simpler' rational functions,

$$R'(t) = \frac{n!}{\prod_{j=0}^{n}(t+j)} = \sum_{k=0}^{n} \frac{A'_k}{t+k}$$

and

$$R''(t) = \frac{(-1)^n \prod_{j=1}^{n}(t-j)}{\prod_{j=0}^{n}(t+j)} = \sum_{k=0}^{n} \frac{A''_k}{t+k}.$$

Then

$$A'_k = \frac{n!}{(-1)^k k!(n-k)!} = (-1)^k \binom{n}{k} \in \mathbb{Z},$$

$$A''_k = \frac{(n+k)!/k!}{(-1)^k k!(n-k)!} = (-1)^k \binom{n}{k}\binom{n+k}{k} \in \mathbb{Z},$$

$$k = 0, 1, \ldots, n,$$

whence

$$R_n(t) = R'(t)R''(t) = \sum_{k=0}^{n} \frac{A'_k A''_k}{(t+k)^2} + \sum_{k=0}^{n}\sum_{\substack{l=0 \\ k \neq l}}^{n} \frac{A'_k A''_l}{(t+k)(t+l)}$$

$$= \sum_{k=0}^{n} \frac{A'_k A''_k}{(t+k)^2} + \sum_{k=0}^{n}\sum_{\substack{l=0 \\ k \neq l}}^{n} \frac{A'_k A''_l}{l-k}\left(\frac{1}{t+k} - \frac{1}{t+l}\right),$$

implying that

$$A_k = A'_k A''_k = \binom{n}{k}^2 \binom{n+k}{k},$$

$$B_k = \sum_{\substack{l=0 \\ l \neq k}}^{n} \frac{A'_k A''_l - A'_l A''_k}{l-k}, \qquad k = 0, 1, \ldots, n.$$

Since $|l - k| \leq n$ in the last sum, the resulting formulae for A_k and B_k give us grounds for the required inclusions. □

Lemma 9.27   *The rational coefficients of the linear form* $r_n = q_n \zeta(2) - p_n$ *satisfy* $q_n \in \mathbb{Z}$ *and* $d_n^2 p_n \in \mathbb{Z}$.

In other words, the sequence

$$q_n = \sum_{k=0}^{n} \binom{n}{k}^2 \binom{n+k}{k}, \qquad n = 0, 1, 2, \ldots,$$

which satisfies the recurrence relation

$$(n + 1)^2 q_{n+1} - (11n^2 + 11n + 3)q_n - n^2 q_{n-1} = 0,$$

is *integer-valued*.

Proof In accordance with the formulae from the proof of Lemma 9.19, we have

$$q_n = \sum_{k=0}^{n} A_k, \qquad p_n = \sum_{k=0}^{n} A_k \sum_{l=1}^{k} \frac{1}{l^2} + \sum_{k=0}^{n} B_k \sum_{l=1}^{k} \frac{1}{l}.$$

Using the inclusions of Lemma 9.26 as well as

$$d_n \sum_{l=1}^{k} \frac{1}{l} \in \mathbb{Z} \quad \text{and} \quad d_n^2 \sum_{l=1}^{k} \frac{1}{l^2} \in \mathbb{Z} \qquad \text{for } k = 0, 1, \dots, n,$$

we arrive at the desired claim. □

Lemma 9.28 *For each $n = 1, 2, \dots,$ the following estimate holds:*

$$0 < |r_n| < \frac{7n}{10^n}.$$

Proof Let us estimate the product $M = m(m+1)\cdots(m+n-1)$ of n successive positive integers. As in the proof of Lemma 2.59, we have

$$\int_{m-1}^{m+n-1} \log x \, dx < \log M = \sum_{l=m}^{m+n-1} \log l < \int_{m}^{m+n} \log x \, dx$$

implying that

$$\log M > (x \log x - x)\Big|_{x=m-1}^{m+n-1} = \log \frac{(m+n-1)^{m+n-1} e^{-n}}{(m-1)^{m-1}},$$

$$\log M < (x \log x - x)\Big|_{x=m}^{m+n} = \log \frac{(m+n)^{m+n} e^{-n}}{m^m}.$$

Thus,

$$n! < \frac{(n+1)^{n+1} e^{-n}}{1^1}, \qquad \prod_{j=1}^{n}(v-j) < \frac{v^v e^{-n}}{(v-n)^{v-n}},$$

$$\prod_{j=1}^{n}(v+j) > \frac{(v+n)^{v+n} e^{-n}}{v^v}.$$

for $v \geq n + 1$; hence

$$0 < (-1)^n R_n(v) < \frac{(n+1)^{n+1}}{v^2 n^n} \frac{n^n v^{3v}}{(v-n)^{v-n}(v+n)^{2(v+n)}}$$

$$= \frac{n}{v^2}\left(1+\frac{1}{n}\right)^{n+1} \frac{(v/n)^{3v}}{(v/n-1)^{v-n}(v/n+1)^{2(v+n)}}$$

$$< \frac{4n}{v^2} f\left(\frac{v}{n}\right),$$

where

$$f(x) = \frac{x^{3x}}{(x-1)^{x-1}(x+1)^{2(x+1)}}.$$

Let C stand for the maximum of the function $f(x)$ in the interval $x > 1$. Then

$$0 < (-1)^n R_n(v) < \frac{4n}{v^2} C^n,$$

implying that

$$0 < (-1)^n r_n < 4nC^n \sum_{v=n+1}^{\infty} \frac{1}{v^2} \leq 4\zeta(2)nC^n < 7nC^n.$$

It remains to compute the maximum C. We have

$$\frac{f'(x)}{f(x)} = \frac{d}{dx}(3x\log x - (x-1)\log(x-1) - 2(x+1)\log(x+1))$$

$$= 3\log x - \log(x-1) - 2\log(x+1) = \log\frac{x^3}{(x-1)(x+1)^2};$$

hence $f'(x) = 0$ if $x^3 = (x-1)(x+1)^2$. A unique root of the latter quadratic equation $-x^2 + x + 1 = 0$ in the interval $x > 1$ is equal to $x_0 = (1+\sqrt{5})/2$. Therefore,

$$C = f(x_0) = \frac{x_0^{3x_0}}{(x_0-1)^{x_0-1}(x_0+1)^{2(x_0+1)}}$$

$$= \frac{x_0-1}{(x_0+1)^2}\left(\frac{x_0^3}{(x_0-1)(x_0+1)^2}\right)^{x_0}$$

$$= \frac{(1+\sqrt{5})/2-1}{((1+\sqrt{5})/2+1)^2} \times 1 = \left(\frac{\sqrt{5}-1}{2}\right)^5 < \frac{1}{10}.$$

This completes our proof of the lemma. □

REMARK 9.29 Dividing both sides of the linear recurrence relation of Theorem 9.21 by n^2, we see that the 'limiting' form of the recurrence for the sequences $\{r_n\}_{n=0}^{\infty}$, $\{q_n\}_{n=0}^{\infty}$ and $\{p_n\}_{n=0}^{\infty}$ is the difference equation

$$r_{n+1} - 11r_n - r_{n-1} = 0$$

with *constant* coefficients. By Theorem 1.27 a general solution of this equation has the form $r_n = c_1 \lambda_1^n + c_2 \lambda_2^n$, where $c_1, c_2 \in \mathbb{R}$, while $\lambda_1 = ((1 - \sqrt{5})/2)^5$ and $\lambda_2 = ((1 + \sqrt{5})/2)^5$ are the roots of characteristic polynomial $\lambda^2 - 11\lambda - 1 = 0$. A consequence of the general formula for r_n is the following limit:

$$\lim_{n \to \infty} \sqrt[n]{|r_n|} = \begin{cases} |\lambda_2| & \text{if } c_2 \neq 0; \\ |\lambda_1| & \text{if } c_2 = 0 \text{ and } c_1 \neq 0. \end{cases}$$

Our original difference equation does not have constant coefficients, and this is an obstacle to obtaining a simple formula for a general solution. However, the limiting relation

$$\limsup_{n \to \infty} \sqrt[n]{|r_n|} \in \{|\lambda_1|, |\lambda_2|\} \tag{9.28}$$

continues to hold whenever r_n is a nontrivial solution. This fact is a classical theorem from analysis due to Poincaré; its proof is not difficult but rather technical [66, Chapter V].

The reasonableness of Poincaré's theorem derives, in part, from its validity for difference equations with constant coefficients (Theorem 1.27).

Elementary estimation shows that $\limsup_{n \to \infty} \sqrt[n]{|r_n|} \leq 1$; hence, using (9.28), we obtain $\limsup_{n \to \infty} \sqrt[n]{|r_n|} = |\lambda_1| < 1/10$. Thus, Poincaré's theorem could save us from the involved computation in the proof of Lemma 9.28.

Proof of Theorem 9.25 Suppose that, on the contrary, $\zeta(2) = a/b$ where a and b are certain positive integers. For each $n = 0, 1, 2, \ldots$, the number

$$r_n^* = bd_n^2 |r_n| = (-1)^n (d_n^2 q_n a - d_n^2 p_n b)$$

is an integer satisfying $0 < r_n^* < 7nbd_n^2 (1/10)^n$. Clearly $r_n \geq 1$, while (9.27) yields $7nbd_n^2 < 3^{2n}$ for all sufficiently large n. The resulting estimate $1 \leq r_n^* < (9/10)^n$ is a contradiction, and proves the theorem. $\qquad\square$

Exercise 9.30 Assume the notation of Exercise 9.24.

(a) Show that, for each $n = 0, 1, \ldots$, at least one of $\tilde{r}_n$ and $\tilde{r}_{n+1}$ is nonzero; in other words, the sequence $\{\tilde{r}_n\}_{n=0}^\infty$ is a *nontrivial* solution of the difference equation from Exercise 9.24(d).

(b) Using Poincaré's theorem, verify that

$$\limsup_{n \to \infty} \sqrt[n]{|\tilde{r}_n|} \in \left\{ (\sqrt{2} - 1)^4, (\sqrt{2} + 1)^4 \right\}.$$

(c) Show, by an elementary estimation, that $|\tilde{r}_n| < Cn$ for a certain constant $C > 0$.

(d) Deduce from (a)–(c) that

$$\limsup_{n\to\infty} \sqrt[n]{|\tilde{r}_n|} = (\sqrt{2} - 1)^4 < \frac{4}{3^3 \times 5}.$$

(e) Show that the coefficients of the linear form $\tilde{r}_n = \tilde{q}_n \zeta(3) - \tilde{p}_n$ satisfy $\tilde{q}_n \in \mathbb{Z}$ and $d_n^3 \tilde{p}_n \in \mathbb{Z}$ for $n = 0, 1, 2, \ldots$

(f) Deduce from (d) and (e) that $\zeta(3)$ is irrational.

Notes

Below this chapter and the rest of this book there lies an iceberg of computation, symbolic and numeric, most of which has not been directly exposed to the reader. The reader would be well advised to keep a computer algebra package open and to implement as much as she or he can! The writers certainly had to use such methods to check or develop many of the more subtle results presented in the book.

One should not miss reading the historical-mathematical account of Apéry's proof [8] of the irrationality of $\zeta(2)$ and $\zeta(3)$ given by Alf van der Poorten in [127]. Our proofs in Sections 9.5 and 9.6 produce Apéry's rational approximations to $\zeta(2)$ (and $\zeta(3)$) but use a somewhat different approach; see [16, 137] for other proofs. It is now apparent that the original construction of Apéry was highly influenced [7, 138] by a continued fraction given by Ramanujan. It should be mentioned that Ramanujan was an indefatigable producer of explicit and highly nontrivial continued fraction expansions, which one could easily classify as beautiful.

Here we limit ourselves to recording the Rogers–Ramanujan continued fractions: for $|q| \le 1$,

$$R(q) = q^{1/5} \prod_{n=0}^{\infty} \frac{(1 - q^{5n+1})(1 - q^{5n+4})}{(1 - q^{5n+2})(1 - q^{5n+3})} = \cfrac{q^{1/5}}{1 + \cfrac{q}{1 + \cfrac{q^2}{1 + \cfrac{q^3}{1 + \ddots}}}}, \qquad (9.29)$$

Then $R(1) = \varphi - 1$, where φ is our old friend the golden ratio, see Exercise 2.31, and $R(q)$ may be thought of as a q-analogue of the golden mean.

Ramanujan showed that $R(e^{-\pi\sqrt{r}})$ is algebraic for each rational number r; Sloane's sequence A082682 in [156] gives the exact values of $r_n = R(e^{-\pi\sqrt{n}})$

for $1 \leq n \leq 10$. In particular, famously,

$$r_2 = \sqrt{\frac{5 + \sqrt{5}}{2} - \frac{1 + \sqrt{5}}{2}}.$$

We highly recommend to our readers that they browse through Berndt's edition of Ramanujan's notebooks [13] and the Andrews–Berndt edition of Ramanujan's 'lost notebook' [5] (see also [6]), since just listing Ramanujan's contributions to this particular subject deserves a separate volume.

Turning to the end of Section 9.4, a recent unifying presentation on the more general Seidel–Stern theorem and its relatives, where terms may be complex, is to be found in [12]. The basic building blocks are as follows, where Z_n represents the nth partial quotient.

Theorem 9.31 (The Seidel–Stern theorem, 1846)   *If each a_n is positive then the sequences Z_{2n} and Z_{2n+1} are monotonic and convergent. If, in addition, $\sum_n a_n$ diverges then Z_n converges.*

Theorem 9.32 (The Stern–Stolz theorem, 1860)   *If Z_n converges then $\sum_n |a_n|$ diverges.*

Theorem 9.33 (Van Vleck's theorem, 1901)   *Suppose that $0 \leq \theta < \pi/2$ and that $|\arg(b_n)| \leq \theta$ whenever $a_n \neq 0$. Then the sequences Z_{2n} and Z_{2n+1} converge. Further, Z_n converges iff $\sum_n |a_n|$ diverges.*

Examples and pictures in [29, 30], and elsewhere, show the need for such restrictions to rule out period-3 and higher-period behaviour of the convergents. They are based on the irregular fraction $S(b)$ given by

$$S(b) = \cfrac{1^2 b_1^2}{1 + \cfrac{2^2 b_2^2}{1 + \cfrac{3^2 b_3^2}{1 + \ddots}}} \tag{9.30}$$

where the string (b_n) is periodic and is most interesting when all terms have the same modulus. The period-2 case is the setting for Ramanujan's AGM fraction. It is convenient to set (9.30) obeys $t_n = q_{n-1}/n!$, where p_n/q_n is the nth partial convergent of $S(b)$, so that

$$t_n = \frac{1}{n} t_{n-1} + \frac{n-1}{n} b_{n-1}^2 t_{n-2}. \tag{9.31}$$

For example, with b_n of period 3 we obtain Figure 9.2 for

$$(b_1, b_2, b_3) = (\exp(i\pi/4), \exp(i\pi/4), \exp(i\pi/4 + 1/\sqrt{2})); \tag{9.32}$$

note the scaling is that suggested by (9.28).

EXERCISE 9.34 (Pictures for $\mathcal{R}$)   In the original Ramanujan fraction setting, draw graphs with $|b_1| = b_2| = 1$ corresponding to that of Figure 9.2. You should see three cases depending on whether none, one or two of the parameters are roots of unity. However, in all cases the graphs produce points lying on two circles and look nothing like Figure 9.2.

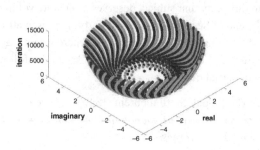

Figure 9.2 Dynamics for cycles of length 3. Shown are the iterates $\sqrt{n}t_n$ for t_n given by (9.31) with the choice (9.32). The odd iterates are light and the even iterates are dark.

We remark that the 'magic' appearance of the identities (9.20) and (9.25) is not accidental: the explicit form of the functions $S_n(t)$ and $\tilde{S}_n(t)$ is the output of the so-called *algorithm of creative telescoping* due to Gosper and Zeilberger, which can be found in [126]. The algorithm is implemented in some computer algebra systems, including Maple^{TM} and Mathematica^R.

It is interesting to note that recurrence equations like (9.22) and (9.26) encode a lot of number theory in addition to the material explored above. In recent years such *Apéry-like* difference equations and their generalisations have become a subject of independent interest [4, 171].

While the irrationality proof for $\zeta(2)$ prefigures that for $\zeta(3)$, the most direct proof of the irrationality of π is probably Ivan Niven's 1947 short proof [120]. It illustrates well the ingredients of many more difficult proofs of the irrationality of other constants and indeed of Lindemann's proof of the transcendence of π, which builds on on Hermite's 1873 proof of the transcendence of e.

Theorem 9.35 ([120])   *The number π is irrational.*

Proof   Let $\pi = a/b$, the quotient of positive integers. We define the polynomials $f(x) = x^n(a - bx)^n/n!$ and

$$F(x) = f(x) - f^{(2)}(x) + f^{(4)}(x) - \cdots + (-1)^n f^{(2n)}(x);$$

the positive integer n will be specified later. Since $n!f(x)$ has integral coefficients and terms in x of degree not less than n, the polynomial $f(x)$ and its derivatives $f^{(j)}(x)$ have integral values for $x = 0$; also for $x = \pi = a/b$, since $f(x) = f(a/b - x)$. By elementary calculus we have

$$\frac{d}{dx}(F'(x)\sin x - F(x)\cos x) = F''(x)\sin x + F(x)\sin x = f(x)\sin x$$

and

$$\int_0^\pi f(x)\sin x \, dx = (F'(x)\sin x - F(x)\cos x)\big|_0^\pi = F(\pi) + F(0). \quad (9.33)$$

Now $F(\pi) + F(0)$ is an *integer*, since $f^{(j)}(0)$ and $f^{(j)}(\pi)$ are integers. But, for $0 < x < \pi$,

$$0 < f(x)\sin x < \frac{\pi^n a^n}{n!},$$

so that the integral in (9.33) is positive but less than 1 for sufficiently large n. Thus (9.33) is false, and so is our assumption that π is rational. □

This proof can be enhanced to cover $\zeta(2)$ as Niven did later [121].

There is a deep connection between the (classical) orthogonal polynomials with respect to a linear functional $\mathcal{L}\colon \mathbb{C}[x] \to \mathbb{C}$ and the continued fraction expansions of the generating function $\sum_{n=0}^\infty \mathcal{L}(x^n)z^n$ for its moments; this has nontrivial applications to the evaluation of Kronecker–Hankel determinants. We refer the interested reader to the highly accessible review [89] of this story (see also [50, pp. 91–99]).

Appendix A

Selected continued fractions

We include a collection of attractive continued fractions both numeric and functional. Where continued fractions have been discussed in the book, corresponding links to the text are provided.

A.1 Regular continued fractions

The majority of examples in this section are computed in the computer algebra system Pari-GP or are to be found in various public sources, such as *Wikipedia* and *MathWorld* at http://mathworld.wolfram.com/.

A selection of quadratic irrationals follows:

$$
\begin{aligned}
&\text{(A.1)} \quad (1 + \sqrt{5})/2 = [\,\overline{1}\,] = [1; 1, 1, 1, 1, \dots] && \text{(Ex. 2.31)};\\
&\text{(A.2)} \quad \sqrt{2} = [1; \overline{2}\,] = [1; 2, 2, 2, 2, \dots];\\
&\text{(A.3)} \quad \sqrt{3} = [1; \overline{1, 2}\,] = [1; 1, 2, 1, 2, \dots];\\
&\text{(A.4)} \quad \sqrt{5} = [2; \overline{4}\,] = [2; 4, 4, 4, 4, \dots];\\
&\text{(A.5)} \quad \sqrt{6} = [2; \overline{2, 4}\,] = [2; 2, 4, 2, 4, \dots];\\
&\text{(A.6)} \quad \sqrt{7} = [2; \overline{1, 1, 1, 4}\,];\\
&\text{(A.7)} \quad \sqrt{8} = [2; \overline{1, 4}\,];\\
&\text{(A.8)} \quad \sqrt{10} = [3; \overline{6}\,];\\
&\text{(A.9)} \quad \sqrt{13} = [3; \overline{1, 1, 1, 1, 6}\,];\\
&\text{(A.10)} \quad \sqrt{14} = [3; \overline{1, 2, 1, 6}\,];\\
&\text{(A.11)} \quad \sqrt{19} = [4; \overline{2, 1, 3, 1, 2, 8}\,];\\
&\text{(A.12)} \quad \sqrt{21} = [4; \overline{1, 1, 2, 1, 1, 8}\,];\\
&\text{(A.13)} \quad \sqrt{22} = [4; \overline{1, 2, 4, 2, 1, 8}\,];\\
&\text{(A.14)} \quad \sqrt{31} = [5; \overline{1, 1, 3, 5, 3, 1, 1, 10}\,];\\
&\text{(A.15)} \quad \sqrt{43} = [6; \overline{1, 1, 3, 1, 5, 1, 3, 1, 1, 12}\,];\\
&\text{(A.16)} \quad \sqrt{46} = [6; \overline{1, 3, 1, 1, 2, 6, 2, 1, 1, 3, 1, 12}\,] && \text{(Example 4.7)};
\end{aligned}
$$

(A.17) $\sqrt{61} = [7; \overline{1,4,3,1,2,2,1,3,4,1,14}\,]$;
(A.18) $\sqrt{76} = [8; \overline{1,2,1,1,5,4,5,1,1,2,1,16}\,]$;
(A.19) $\sqrt{94} = [9; \overline{1,2,3,1,1,5,1,8,1,5,1,1,3,2,1,18}\,]$.

In the above expressions the vinculum (overbar) denotes periodic repetition of the corresponding part (see Section 2.9).

Some 'arithmetic-progression' continued fractions (see also [54, 88] and [159, Theorem 2]) are the following:

(A.20) $e = [2; (1,2k,1)_{k=1}^{\infty}] = [2; 1,2,1,1,4,1,1,6,1,1,8,\dots]$ (Section 2.12);
(A.21) $e^{1/n} = [1; n-1,1,1,1,3n-1,1,1,1,5n-1,1,1,1,7n-1,1,1,\dots]$ (Ex. 2.54);
(A.22) $e^{2/(2n+1)} = [(1, 3k(2n+1)+n, 6(2k+1)(2n+1),$
$\qquad\qquad (3k+2)(2n+1)+n, 1)_{k=0}^{\infty}]$ (Ex. 2.57);
(A.23) $e^2 = [7; 2,1,1,3,18,5,1,1,6,30,8,1,1,9,42,11,1,1,12,54,14,\dots]$;
(A.24) $2e = [5; 2,3,(2k,3,1,2k)_{k=1}^{\infty}]$;
(A.25) $3e = [8; 6,2,5,(2k,5,1,2k,5,1,2k,1)_{k=1}^{\infty}]$;
(A.26) $4e = [[0; 1,6,1,7,2,(7,k+1,7,1,k,1)_{k=1}^{\infty}]$;
(A.27) $e/2 = [1; 2,(2k+1,3,1,2k+1,1,3)_{k=0}^{\infty}]$;
(A.28) $e/3 = [0; 1,9,(1,1,2k+1,5,1,2k+1,1,1,18k+26)_{k=0}^{\infty}]$;
(A.29) $e/4 = [0; 1,2,8,3,(1,1,1,k,7,1,k,2)_{k=1}^{\infty}]$;
(A.30) $ne^{1/n} = [n+1; (2n-1,2k,1)_{k=1}^{\infty}]$;
(A.31) $e^{1/n}/n = [0; n-1,2n,(1,2k,2n-1)_{k=1}^{\infty}]$;
(A.32) $\tan(1/n) = [0; n-1,1,3n-2,1,5n-2,1,7n-2,1,9n-2,1,\dots]$,
     with a special case for $n=1$ (Ex. 2.56);
(A.33) $\tan(1) = [1; (2k+1,1)_{k=0}^{\infty}]$
$\qquad\qquad = [1; 1,1,3,1,5,1,7,1,9,1,11,1,13,1,15,1,17,1,19,1,\dots]$.

For the modified, or hyperbolic, Bessel function of the first kind,

$$I_c(x) = \left(\frac{x}{2}\right)^c \sum_{n=0}^{\infty} \frac{(x^2/4)^n}{\Gamma(c+n+1)\,n!}$$

(compare with the function $f(c,x)$ in Section 2.10), we define the following function on the rationals $c = a/b$:

$$S(a,b) = \frac{I_{a/b}(2/b)}{I_{a/b+1}(2/b)},$$

with a and b in lowest terms. Then for all nonnegative rationals we have

(A.34) $S(a,b) = [a+b; a+2b, a+3b, a+4b, \cdots]$,

with similar formulae for negative rationals; in particular, for the case $a/b = 0/1$

we have

(A.35) $S(0,1) = I_0(2)/I_1(2) = [1; 2, 3, 4, 5, 6, 7, \ldots]$.

We also have 'geometric-progression' continued fractions: for integer $a > 1$,

(A.36) $\displaystyle\prod_{k=0}^{\infty} \frac{(1 - a^{-(5k+2)})(1 - a^{-(5k+3)})}{(1 - a^{-(5k+1)})(1 - a^{-(5k+4)})} = [1; (a^k, a^k)_{k=1}^{\infty}]$

$$= [1; a, a, a^2, a^2, a^3, a^3, a^4, a^4, a^5, \ldots];$$

(A.37) $\displaystyle\prod_{k=0}^{\infty} \frac{1 - a^{-(3k+2)}}{1 - a^{-(3k+1)}} = [1; (a^k - 1, 1)_{k=1}^{\infty}]$

$$= [1; a - 1, 1, a^2 - 1, 1, a^3 - 1, 1, a^4 - 1, 1, \ldots].$$

The first instance follows from the Rogers–Ramanujan continued fraction (9.29), while the second is equivalent to the irregular continued fraction (A.75) given below.

For the majority of mathematical constants [58], such as π, the logarithms of rationals, Euler's constant γ, Catalan's constant G and the values of Riemann's zeta function $\zeta(s)$ for positive integers, no pattern in partial quotients has ever been detected. Many of these constants have not even been proved to be irrational. Here we reproduce some examples:

(A.38) $\pi = 3.141\,592\,653\,589\,793\,238\,462\,643\,383\,279\,502\,884\,197\,1\ldots$
$= [3; 7, 15, 1, 292, 1, 1, 1, 2, 1, 3, 1, 14, 2, 1, 1, 2, 2, 2, 2, 1, 84, 2, \ldots]$;

(A.39) $\gamma = 0.577\,215\,664\,901\,532\,860\,606\,512\,090\,082\,402\,431\,042\,1\ldots$
$= [0; 1, 1, 2, 1, 2, 1, 4, 3, 13, 5, 1, 1, 8, 1, 2, 4, 1, 1, 40, 1, 11, 3, \ldots]$;

(A.40) $\log 2 = 0.693\,147\,180\,559\,945\,309\,417\,232\,121\,458\,176\,568\,075\,5\ldots$
$= [0; 1, 2, 3, 1, 6, 3, 1, 1, 2, 1, 1, 1, 1, 3, 10, 1, 1, 1, 2, 1, 1, 1, 1, \ldots]$;

(A.41) $\zeta(2) = \pi^2/6$
$= 1.644\,934\,066\,848\,226\,436\,472\,415\,166\,646\,025\,189\,218\,9\ldots$
$= [1; 1, 1, 1, 4, 2, 4, 7, 1, 4, 2, 3, 4, 10, 1, 2, 1, 1, 1, 15, 1, 3, 6, \ldots]$;

(A.42) $G = 0.915\,965\,594\,177\,219\,015\,054\,603\,514\,932\,384\,110\,774\,1\ldots$
$= [0; 1, 10, 1, 8, 1, 88, 4, 1, 1, 7, 22, 1, 2, 3, 26, 1, 11, 1, 10, 1, 9, \ldots]$;

(A.43) $\zeta(3) = 1.202\,056\,903\,159\,594\,285\,399\,738\,161\,511\,449\,990\,764\,9\ldots$
$= [1; 4, 1, 18, 1, 1, 1, 4, 1, 9, 9, 2, 1, 1, 1, 2, 7, 1, 1, 7, 11, 1, 1, \ldots]$;

(A.44) $\zeta(5) = 1.036\,927\,755\,143\,369\,926\,331\,365\,486\,457\,034\,168\,057\,0\ldots$
$= [1; 27, 12, 1, 1, 15, 1, 5, 1, 2, 19, 1, 1, 32, 1, 13, 1, 1, 1, 3, 1, 3, \ldots]$.

A.2 Irregular continued fractions

In this section we employ the notation

$$a_0 + \frac{b_1}{|a_1|} + \frac{b_2}{|a_2|} + \cdots + \frac{b_n}{|a_n|} + \cdots = a_0 + \cfrac{b_1}{a_1 + \cfrac{b_2}{a_2 + \cfrac{\ddots}{\;\; + a_{n-1} + \cfrac{b_n}{a_n + \cdot_{\cdot_\cdot}}}}}$$

for irregular continued fractions (from Chapter 9). Variations such as

$$\frac{b_1}{|a_1|} - \frac{b_2}{|a_2|} - \frac{b_3}{|a_3|} - \cdots = \cfrac{b_1}{a_1 - \cfrac{b_2}{a_2 - \cfrac{b_3}{a_3 - \cdot_{\cdot_\cdot}}}}$$

(instead of negating the sign of the b_n) can be used for aesthetic purposes.

The principal source for this section is the *Digital library of mathematical functions* [123], but we also acknowledge [5, 13, 50, 78, 103] amongst other references.

A.2.1 Hypergeometric functions

A generic hypergeometric series

$$F(a, b; c; z) = \sum_{n=0}^{\infty} \frac{(a)_n (b)_n}{n! \, (c)_n} z^n,$$

where $(a)_n = \Gamma(a + n)/\Gamma(a) = a(a + 1) \cdots (a + n - 1)$ denotes the *Pochhammer symbol* (or *shifted factorial*), converges inside the unit disc. The second-order homogeneous linear differential equation satisfied by the series (see Section 9.3) can be used for its analytical continuation to a z-holomorphic function on $\mathbb{C} \setminus [1, \infty)$ (equivalently, $|\arg(1 - z)| < \pi$); it is this function, also denoted by $F(a, b; c; z)$, that goes by the name *hypergeometric*. All elementary functions are known to be special or limiting cases of the hypergeometric function; we illustrate this in Section 9.3 and use it (implicitly) in Subsection A.2.2 below.

If $z \in \mathbb{C} \setminus [1, \infty)$ then we have the Gauss continued fraction (Theorem 9.10)

$$\text{(A.45)} \quad \frac{c \, F(a, b; c; z)}{F(a + 1, b; c + 1; z)} = c + \frac{\lambda_0 z}{|c + 1|} + \frac{\lambda_1 z}{|c + 2|} + \cdots + \frac{\lambda_n z}{|c + n + 1|} + \cdots,$$

where $\lambda_{2k-1} = (a + k)(b - c - k)$ and $\lambda_{2k} = (a - c - k)(b + k)$. With the help of contiguous relations we also have, for $\mathrm{Re}\, z < 1/2$,

(A.46) $\dfrac{F(a,b;c;z)}{F(a+1,b+1;c+1;z)} = a_0 + \dfrac{b_1|}{|a_1} + \dfrac{b_2|}{|a_2} + \cdots + \dfrac{b_n|}{|a_n} + \cdots,$

where $a_n = c + n - (a + b + 2n + 1)z$ and $b_n = (a + n)(b + n)z(1 - z)$; see also [50, pp. 295–309]. Furthermore,

(A.47) $\displaystyle\sum_{n=0}^{\infty} \dfrac{(-z)^n}{(c)_{n+1}} = \dfrac{1|}{|c} + \dfrac{z|}{|1} + \dfrac{1|}{|c} + \dfrac{z|}{|1} + \dfrac{2|}{|c} + \dfrac{z|}{|1} + \dfrac{3|}{|c} + \dfrac{z|}{|1} + \cdots + \dfrac{n|}{|c} + \dfrac{z|}{|1} + \cdots$

(A.48) $\qquad\qquad = \dfrac{1|}{|c+z} - \dfrac{z|}{|c+z+1} - \dfrac{2z|}{|c+z+2} - \cdots - \dfrac{nz|}{|c+z+n} - \cdots$

and

(A.49) $\dfrac{a\mathcal{F}(a/b; z/b^2)}{\mathcal{F}(a/b+1; z/b^2)} = a + \dfrac{z|}{|a+b} + \dfrac{z|}{|a+2b} + \cdots + \dfrac{z|}{|a+nb} + \cdots$

(compare with (A.34)), where

$$\mathcal{F}(a;z) = \sum_{n=0}^{\infty} \dfrac{z^n}{n!\,(a)_n} = \lim_{c\to\infty} F(c,c;a;z/c^2),$$

which function is clearly related to the modified Bessel function of the first kind.

The confluent hypergeometric function

$$M(a,b;z) = \sum_{n=0}^{\infty} \dfrac{(a)_n z^n}{n!\,(b)_n} = \lim_{c\to\infty} F(a,c;b;z/c)$$

satisfies Kummer's differential equation

$$z\dfrac{\mathrm{d}^2 M}{\mathrm{d}z^2} + (b - z)\dfrac{\mathrm{d}M}{\mathrm{d}z} - aM = 0.$$

Another solution to this equation,

$$U(a,b;z) = \dfrac{\Gamma(1-b)}{\Gamma(a-b+1)}\,M(a,b;z) + \dfrac{\Gamma(b-1)}{\Gamma(a)}\,z^{1-b}M(a-b+1, 2-b; z)$$

(for $b = N$, an integer, one replaces the right-hand side by its limit as $b \to N$), is known as the Tricomi confluent hypergeometric function. Then

(A.50) $\dfrac{bM(a,b;z)}{M(a+1,b+1;z)} = b + \dfrac{v_0 z|}{|b+1} + \dfrac{v_1 z|}{|b+2} + \cdots + \dfrac{v_n z|}{|b+n+1} + \cdots,$

where $v_{2k-1} = a + k$ and $v_{2k} = a - b - k$, and

(A.51) $\dfrac{U(a, b+1; z)}{U(a, b; z)} = 1 + \dfrac{\hat{v}_0/z}{\vert 1} + \dfrac{\hat{v}_1/z}{\vert 1} + \cdots + \dfrac{\hat{v}_n/z}{\vert 1} + \cdots,$

where $v_{2k-1} = a - b + k$ and $v_{2k} = a + k$. Furthermore, for $\mathrm{Re}\, z > 0$,

(A.52) $\operatorname{erf} z = 1 - \dfrac{ze^{-z^2}/\sqrt{\pi}}{\vert z^2} + \dfrac{1/2}{\vert 1} + \dfrac{1}{\vert z^2} + \dfrac{3/2}{\vert 1} + \cdots + \dfrac{n}{\vert z^2} + \dfrac{n+1/2}{\vert 1} + \cdots$

(A.53) $= 1 - \dfrac{2ze^{-z^2}/\sqrt{\pi}}{\vert 2z^2 + 1} - \dfrac{1 \times 2}{\vert 2z^2 + 5} - \dfrac{3 \times 4}{\vert 2z^2 + 9} + \cdots + \dfrac{(2n-1)(2n)}{\vert 2z^2 + 4n + 1} + \cdots$

(A.54) $= 1 - \dfrac{2e^{-z^2}/\sqrt{\pi}}{\vert 2z} + \dfrac{1}{\vert z} + \dfrac{2}{\vert 2z} + \dfrac{3}{\vert z} + \dfrac{4}{\vert 2z} + \cdots + \dfrac{2n-1}{\vert z} + \dfrac{2n}{\vert 2z} + \cdots$

where the Gauss error function $\operatorname{erf} z$ is given by

$$\operatorname{erf} z = \frac{2}{\sqrt{\pi}} \int_0^z e^{-x^2} \mathrm{d}x = \frac{2}{\sqrt{\pi}} \sum_{n=0}^{\infty} \frac{(-1)^n z^{2n+1}}{n!\,(2n+1)}.$$

See also [44, pp. 255–260, 263–267, 270–273] for related continued fractions.

Some limiting cases of the Gauss continued fraction (A.45) correspond to continued fraction expansions involving incomplete gamma and beta functions (see, for example, [78] and [50, pp. 240–251]). We limit ourselves here to presenting a different continued fraction for the gamma function itself, which is useful even though it does not have an explicit pattern: for $\mathrm{Re}\, z > 0$,

(A.55) $\log \Gamma(z) + z - (z - 1/2)\log z - (1/2)\log(2\pi)$

$\qquad = \dfrac{a_0}{\vert z} + \dfrac{a_1}{\vert z} + \dfrac{a_2}{\vert z} + \dfrac{a_3}{\vert z} + \dfrac{a_4}{\vert z} + \dfrac{a_5}{\vert z} + \dfrac{a_6}{\vert z} + \cdots,$

where

$$a_0 = \tfrac{1}{12}, \quad a_1 = \tfrac{1}{30}, \quad a_2 = \tfrac{53}{210}, \quad a_3 = \tfrac{195}{371},$$
$$a_4 = \tfrac{22\,999}{22\,737}, \quad a_5 = \tfrac{29\,944\,523}{19\,733\,142}, \quad a_6 = \tfrac{109\,535\,241\,009}{48\,264\,275\,462}.$$

For exact values of a_7 to a_{11} and approximate values up to a_{40} see [44] (also [50, pp. 223–228] and [78, pp. 348–35]).

A.2.2 Elementary functions

Several continued fractions involving logarithms are given in [103, pp. 566–568] (see also [50, pp. 196–200]); here we limit ourselves to the following examples:

(A.56) $\log(1+z) = \dfrac{z}{\lvert 1} + \dfrac{z}{\lvert 2} + \dfrac{z}{\lvert 3} + \dfrac{2^2 z}{\lvert 4} + \dfrac{2^2 z}{\lvert 5} + \cdots + \dfrac{n^2 z}{\lvert 2n} + \dfrac{n^2 z}{\lvert 2n+1} + \cdots$;

(A.57) $\log(1+z) - \displaystyle\sum_{n=1}^{N-1} \dfrac{(-1)^{n-1} z^n}{n} = \dfrac{(-1)^{N-1} z^N}{\lvert N} + \dfrac{N^2 z}{\lvert N+1} + \dfrac{1^2 z}{\lvert N+2}$

$$+ \dfrac{(N+1)^2 z}{\lvert N+3} + \dfrac{2^2 z}{\lvert N+4} + \dfrac{(N+2)^2 z}{\lvert N+5} + \cdots$$

$$+ \dfrac{n^2 z}{\lvert N+2n} + \dfrac{(N+n)^2 z}{\lvert N+2n+1} + \cdots$$

(Section 9.3);

valid for $z \in \mathbb{C} \setminus (-\infty, -1]$

(A.58) $\log \dfrac{1+z}{1-z} = \dfrac{2z}{\lvert 1} - \dfrac{z^2}{\lvert 3+z^2} - \dfrac{3^2 z^2}{\lvert 5+3z^2} - \cdots - \dfrac{(2n-1)^2 z^2}{\lvert (2n+1)+(2n-1)z^2} - \cdots$

(Section 9.2);

(A.59) $= \dfrac{2z}{\lvert 1} - \dfrac{z^2}{\lvert 3} - \dfrac{2^2 z^2}{\lvert 5} - \dfrac{3^2 z^2}{\lvert 7} - \cdots - \dfrac{n^2 z}{\lvert 2n+1} - \cdots$

valid for $z \in \mathbb{C} \setminus (-\infty, -1] \cup [1, \infty)$.

Next come continued fraction expansions of the exponential function (see also [103, pp. 563–564] and [50, pp. 193–195]): for $z \in \mathbb{C}$,

(A.60) $e^z = \dfrac{1}{\lvert 1} - \dfrac{z}{\lvert 1} + \dfrac{z}{\lvert 2} - \dfrac{z}{\lvert 3} + \dfrac{z}{\lvert 2} - \dfrac{z}{\lvert 5} + \cdots + \dfrac{z}{\lvert 2} - \dfrac{z}{\lvert 2n+1} + \cdots$

(A.61) $= 1 + \dfrac{z}{\lvert 1} - \dfrac{z}{\lvert 2} + \dfrac{z}{\lvert 3} - \dfrac{z}{\lvert 2} + \dfrac{z}{\lvert 5} - \cdots - \dfrac{z}{\lvert 2} + \dfrac{z}{\lvert 2n+1} - \cdots$;

(A.62) $e^{2z} = 1 + \dfrac{2z}{\lvert 1-z} + \dfrac{z^2/3}{\lvert 1} + \dfrac{z^2/15}{\lvert 1} + \dfrac{z^2/35}{\lvert 1} + \cdots + \dfrac{z^2/(4n^2-1)}{\lvert 1} + \cdots$;

(A.63) $\dfrac{e^z - 1}{e^z + 1} = \dfrac{z}{\lvert 2} + \dfrac{z^2}{\lvert 6} + \dfrac{z^2}{\lvert 10} + \dfrac{z^2}{\lvert 14} + \cdots + \dfrac{z^2}{\lvert 4n+2} + \cdots$;

(A.64) $e^z - \displaystyle\sum_{n=0}^{N-1} \dfrac{z^n}{n!} = \dfrac{z^N}{\lvert N!} - \dfrac{N! z}{\lvert N+1} + \dfrac{z}{\lvert N+2} - \dfrac{(N+1)z}{\lvert N+3} + \dfrac{2z}{\lvert N+4} - \dfrac{(N+2)z}{\lvert N+5}$

$$+ \cdots + \dfrac{nz}{\lvert N+2n} - \dfrac{(N+n)z}{\lvert N+2n+1} + \cdots$$.

In addition we have, for $z \in \mathbb{C} \setminus [-i, i]$,

$$(\text{A.65}) \quad e^{2a \arctan(1/z)} = 1 + \frac{2a}{\lvert z - a} + \frac{a^2 + 1}{\lvert 3z} + \frac{a^2 + 4}{\lvert 5z} + \cdots + \frac{a^2 + n^2}{\lvert (2n+1)z} + \cdots$$

as well as expansions for inverse trigonometric functions:

$$(\text{A.66}) \quad \arctan z = \frac{z}{\lvert 1} + \frac{z^2}{\lvert 3 - z^2} + \frac{3^2 z^2}{\lvert 5 - 3z^2} + \cdots + \frac{(2n-1)^2 z^2}{\lvert (2n+1) - (2n-1)z^2} + \cdots$$

$$(\text{A.67}) \quad = \frac{z}{\lvert 1} + \frac{z^2}{\lvert 3} + \frac{2^2 z^2}{\lvert 5} + \frac{3^2 z^2}{\lvert 7} + \cdots + \frac{n^2 z^2}{\lvert 2n+1} + \cdots \, ; \qquad (\text{Section 9.2})$$

$$(\text{A.68}) \quad \arctan z - \sum_{n=0}^{N-1} \frac{(-1)^n z^{2n+1}}{2n+1} = \frac{(-1)^N z^{2N+1}}{\lvert 2N+1} + \frac{(2N+1)^2 z^2}{\lvert 2N+3} + \frac{2^2 z^2}{\lvert 2N+5}$$

$$+ \frac{(2N+3)^2 z^2}{\lvert 2N+7} + \frac{4^2 z^2}{\lvert 2N+9} + \cdots$$

$$+ \frac{(2N+2n-1)^2 z^2}{\lvert 2N+4n-1} + \frac{(2n)^2 z^2}{\lvert 2N+4n+1} + \cdots$$

$$(\text{Section 9.3})$$

(the latter three expansions are valid for $z \in \mathbb{C} \setminus (-i\infty, -i] \cup [i, i\infty)$);

$$(\text{A.69}) \quad \frac{\arcsin z}{\sqrt{1 - z^2}} = \frac{z}{\lvert 1} - \frac{1 \times 2z^2}{\lvert 3} - \frac{1 \times 2z^2}{\lvert 5} - \frac{3 \times 4z^2}{\lvert 7} - \frac{3 \times 4z^2}{\lvert 9} - \cdots$$

$$- \frac{(2n-1)(2n)z^2}{\lvert 4n-1} - \frac{(2n-1)(2n)z^2}{\lvert 4n+1} - \cdots ,$$

valid for $z \in \mathbb{C} \setminus (-\infty, -1] \cup [1, \infty)$; and also

$$(\text{A.70}) \quad \arctan z = \frac{z}{\lvert 1} + \frac{z^2}{\lvert 1} + \frac{2(1 + z^2)}{\lvert 1} + \frac{3z^2}{\lvert 1} + \frac{4(1 + z^2)}{\lvert 1} + \cdots$$

$$+ \frac{(2n-1)z^2}{\lvert 1} + \frac{2n(1 + z^2)}{\lvert 1} + \cdots ,$$

$$(\text{A.71}) \quad \frac{\arcsin z}{\sqrt{1 - z^2}} = \frac{z}{\lvert 1} + \frac{-2z^2}{\lvert 1} + \frac{2(1 - z^2)}{\lvert 1} + \frac{-4z^2}{\lvert 1} + \frac{4(1 - z^2)}{\lvert 1} + \cdots$$

$$+ \frac{-2nz^2}{\lvert 1} + \frac{2n(1 - z^2)}{\lvert 1} + \cdots ,$$

valid for $\operatorname{Re} z^2 > -1/2$ and $\operatorname{Re} z^2 < 1/2$, respectively. See [78, pp. 560–571] and [50, pp. 201–203, 205–210] for other continued fractions involving inverse trigonometric functions. Inverse hyperbolic functions are related to these by the formulae $\operatorname{arcsinh}(iz) = i \arcsin z$ and $\operatorname{arctanh}(iz) = i \arctan z$. Their continued fractions are discussed in [103, pp. 569–571]; see also [50, pp. 211–217].

Finally, we reproduce two examples related to the tangent function:

$$(A.72) \quad \tan z = \cfrac{z|}{|1} - \cfrac{z^2|}{|3} - \cfrac{z^2|}{|5} - \cdots - \cfrac{z^2|}{|2n+1} - \cdots$$

for $z \notin \pi/2 + \pi\mathbb{Z}$; and

$$(A.73) \quad \tan az = \cfrac{a \tan z|}{|1} + \cfrac{(1-a^2)\tan^2 z|}{|3} + \cfrac{(4-a^2)\tan^2 z|}{|5} + \cdots$$

$$+ \cfrac{(n^2 - a^2)\tan^2 z|}{|2n+1} + \cdots$$

for $z \notin \pi/2 + \pi\mathbb{Z}$ and $|\operatorname{Re} z| < \pi/2$.

A.2.3 q-Series

In this part we assume $|q| < 1$ and implement the q-Pochhammer notation,

$$(a; q)_n = \prod_{k=0}^{n-1} (1 - aq^k) \quad \text{and} \quad (a; q)_\infty = \lim_{n \to \infty} (a; q)_n = \prod_{k=0}^{\infty} (1 - aq^k).$$

Then the Rogers–Ramanujan continued fraction (9.29) can be expressed as follows:

$$(A.74) \quad \frac{(q; q^5)_\infty (q^4; q^5)_\infty}{(q^2; q^5)_\infty (q^3; q^5)_\infty} = \cfrac{1|}{|1} + \cfrac{q|}{|1} + \cfrac{q^2|}{|1} + \cfrac{q^3|}{|1} + \cdots + \cfrac{q^n|}{|1} + \cdots.$$

We also have:

$$(A.75) \quad \frac{(q^2; q^3)_\infty}{(q; q^3)_\infty} = \cfrac{1|}{|1} - \cfrac{q|}{|1+q} - \cfrac{q^3|}{|1+q^2} - \cfrac{q^5|}{|1+q^3} - \cdots - \cfrac{q^{2n-1}|}{|1+q^n} - \cdots;$$

$$(A.76) \quad \frac{(q; q^2)_\infty}{(q^3; q^6)_\infty^3} = \cfrac{1|}{|1} + \cfrac{q+q^2|}{|1} + \cfrac{q^2+q^4|}{|1} + \cfrac{q^3+q^6|}{|1} + \cdots + \cfrac{q^n+q^{2n}|}{|1} + \cdots;$$

$$(A.77) \quad 1 - \sum_{n=0}^{\infty} q^{n(3n-1)/2}(1 - q^n) = \cfrac{2|}{|2} + \cfrac{q+q|}{|1} + \cfrac{q^2+q^3|}{|1}$$

$$+ \cfrac{q^3+q^5|}{|1} + \cdots + \cfrac{q^n+q^{2n-1}|}{|1} + \cdots;$$

$$(A.78) \quad \frac{(-aq^2; q^2)_\infty}{(-aq; q^2)_\infty} = \cfrac{1|}{|1} + \cfrac{aq|}{|1} + \cfrac{q+aq^2|}{|1} + \cfrac{aq^3|}{|1} + \cfrac{q^2+aq^4|}{|1}$$

$$+ \cfrac{aq^5|}{|1} + \cfrac{q^3+aq^6|}{|1} + \cdots \qquad (a \notin -q^{-2\mathbb{Z}_{>0}+1});$$

$$\text{(A.79)} \quad \sum_{n=0}^{\infty}(-a)^n q^{n(n+1)/2} = \cfrac{1|}{|1} + \cfrac{aq|}{|1} + \cfrac{a(q^2-q)|}{|1} + \cfrac{aq^3|}{|1} + \cfrac{a(q^4-q^2)|}{|1}$$

$$+ \cfrac{aq^5|}{|1} + \cfrac{a(q^6-q^3)|}{|1} + \cdots .$$

More generally, for complex a, b and λ,

$$\text{(A.80)} \quad \frac{G(aq,b,\lambda q)}{G(a,b,\lambda)} = \cfrac{1|}{|1} + \cfrac{aq+\lambda q|}{|1} + \cfrac{bq+\lambda q^2|}{|1} + \cfrac{aq^2+\lambda q^3|}{|1} + \cfrac{bq^2+\lambda q^4|}{|1}$$

$$+ \cfrac{aq^3+\lambda q^5|}{|1} + \cfrac{bq^3+\lambda q^6|}{|1} + \cdots$$

where

$$G(a,b,\lambda) = \sum_{n=0}^{\infty} \frac{a^n(-\lambda/a;q)_n q^{n(n+1)/2}}{(q;q)_n(-bq;q)_n}.$$

The limiting case,

$$g(b,\lambda) = \lim_{a \to 0} G(a,b,\lambda) = \sum_{n=0}^{\infty} \frac{\lambda^n q^{n^2}}{(q;q)_n(-bq;q)_n},$$

leads to different continued fraction expansions:

$$\text{(A.81)} \quad \frac{g(b,\lambda q)}{g(b,\lambda)} = \cfrac{1|}{|1} + \cfrac{\lambda q|}{|1} + \cfrac{\lambda q^2+bq|}{|1} + \cfrac{\lambda q^3|}{|1} + \cfrac{\lambda q^4+bq^2|}{|1} + \cfrac{\lambda q^5|}{|1} + \cdots$$

$$\text{(A.82)} \qquad = \cfrac{1|}{|1} + \cfrac{\lambda q|}{|1+bq} + \cfrac{\lambda q^2|}{|1+bq^2} + \cfrac{\lambda q^3|}{|1+bq^3} + \cfrac{\lambda q^4|}{|1+bq^4} + \cdots$$

$$\text{(A.83)} \qquad = \cfrac{1|}{|1-b} + \cfrac{b+\lambda q|}{|1-b} + \cfrac{b+\lambda q^2|}{|1-b} + \cfrac{b+\lambda q^3|}{|1-b} + \cfrac{b+\lambda q^4|}{|1-b} + \cdots .$$

Similar, though more involved, expansions are available for (A.80) as well.

For other examples of q-series continued fractions we refer the reader to the original source [5, Part I].

A.2.4 Miscellaneous continued fractions

For $\operatorname{Re} a > 0$,

$$\text{(A.84)} \quad 2\sum_{n=0}^{\infty}\frac{(-1)^n}{a+2n+1} = \cfrac{1|}{|a} + \cfrac{1^2|}{|a} + \cfrac{2^2|}{|a} + \cfrac{3^2|}{|a} + \cdots + \cfrac{n^2|}{|a} + \cdots \qquad \text{(Section 9.4)};$$

$$\text{(A.85)} \quad 1 + 2a\sum_{n=1}^{\infty}\frac{(-1)^n}{a+2n} = \cfrac{1|}{|a} + \cfrac{1\times 2|}{|a} + \cfrac{2\times 3|}{|a} + \cfrac{3\times 4|}{|a} + \cdots + \cfrac{n(n+1)|}{|a} + \cdots ;$$

(A.86) $\displaystyle 2\sum_{n=0}^{\infty}\frac{1}{(a+2n+1)^2}=\cfrac{1}{a}+\cfrac{1^4}{3a}+\cfrac{2^4}{5a}+\cfrac{3^4}{7a}+\cdots+\cfrac{n^4}{(2n+1)a}+\cdots\,;$

(A.87) $\displaystyle 2\sum_{n=0}^{\infty}\frac{(-1)^n}{(a+2n+1)^2}=\cfrac{1}{a^2-1}+\cfrac{2^2}{1}+\cfrac{2^2}{a^2-1}+\cfrac{4^2}{1}+\cfrac{4^2}{a^2-1}+\cdots$

$$+\cfrac{(2n)^2}{1}+\cfrac{(2n)^2}{a^2-1}+\cdots\,;$$

(A.88) $\displaystyle 1+2a^2\sum_{n=1}^{\infty}\frac{(-1)^n}{(a+n)^2}=\cfrac{1}{a}+\cfrac{1^2}{a}+\cfrac{1\times 2}{a}+\cfrac{2^2}{a}+\cfrac{2\times 3}{a}+\cfrac{3^2}{a}+\cdots$

$$+\cfrac{n^2}{a}+\cfrac{n(n+1)}{a}+\cdots\,;$$

(A.89) $\displaystyle \sum_{n=1}^{\infty}\frac{1}{(a+n)^3}=\cfrac{1}{2a(a+1)}+\cfrac{1^3}{1}+\cfrac{1^3}{6a(a+1)}+\cfrac{2^3}{1}+\cfrac{2^3}{10a(a+1)}+\cdots$

$$+\cfrac{n^3}{1}+\cfrac{n^3}{(4n+2)a(a+1)}+\cdots$$

(A.90) $\displaystyle =\cfrac{1}{2a^2+2a+1}-\cfrac{1^6}{3(2a^2+2a+3)}-\cfrac{2^6}{5(2a^2+2a+7)}-\cdots$

$$-\cfrac{n^6}{(2n+1)(2a^2+2a+n^2+n+1)}-\cdots\,.$$

Specialising these (and some earlier) identities results in

(A.91) $\displaystyle \frac{\pi}{4}=\cfrac{1}{1}+\cfrac{1}{2}+\cfrac{3^2}{2}+\cfrac{5^2}{2}+\cdots+\cfrac{(2n+1)^2}{2}+\cdots$ (Eq. (9.8))

(A.92) $\displaystyle =\cfrac{1}{1}+\cfrac{1^2}{3}+\cfrac{2^2}{5}+\cfrac{3^2}{7}+\cdots+\cfrac{n^2}{2n+1}+\cdots$ (Eq. (9.16));

(A.93) $\displaystyle \frac{\pi^2}{6}=\cfrac{2}{1}+\cfrac{1^4}{3}+\cfrac{2^4}{5}+\cfrac{3^4}{7}+\cdots+\cfrac{n^4}{2n+1}+\cdots$

(A.94) $\displaystyle =1+\cfrac{1}{1}+\cfrac{1^2}{1}+\cfrac{1\times 2}{1}+\cfrac{2^2}{1}+\cfrac{2\times 3}{1}+\cfrac{3^2}{1}+\cdots$

$$+\cfrac{n^2}{1}+\cfrac{n(n+1)}{1}+\cdots$$

(A.95) $\displaystyle =\cfrac{5}{3}+\cfrac{1^4}{25}+\cfrac{2^4}{69}+\cdots+\cfrac{n^4}{11n^2+11n+3}+\cdots$ (Theorem 9.23);

$$\text{(A.96)} \quad 2G = 1 - \cfrac{1}{|3} + \cfrac{2^2}{|1} + \cfrac{2^2}{|3} + \cfrac{4^2}{|1} + \cfrac{4^2}{|3} + \cdots + \cfrac{(2n)^2}{|1} + \cfrac{(2n)^2}{|3} + \cdots$$

$$\text{(A.97)} \quad = 1 + \cfrac{1}{|1/2} + \cfrac{1^2}{|1/2} + \cfrac{1 \times 2}{|1/2} + \cfrac{2^2}{|1/2} + \cfrac{2 \times 3}{|1/2} + \cfrac{3^2}{|1/2} + \cdots$$

$$+ \cfrac{n^2}{|1/2} + \cfrac{n(n+1)}{|1/2} + \cdots \, ;$$

$$\text{(A.98)} \quad \zeta(3) = 1 + \cfrac{1}{|4} + \cfrac{1^3}{|1} + \cfrac{1^3}{|12} + \cfrac{2^3}{|1} + \cfrac{2^3}{|20} + \cdots + \cfrac{n^3}{|1} + \cfrac{n^3}{|4(2n+1)} + \cdots$$

$$\text{(A.99)} \quad = \cfrac{1}{|1} - \cfrac{1^6}{|3 \times 3} - \cfrac{2^6}{|5 \times 7} - \cfrac{3^6}{|7 \times 13} - \cdots$$

$$- \cfrac{n^6}{|(2n+1)(n^2 + n + 1)} - \cdots$$

$$\text{(A.100)} \quad = \cfrac{6}{|5} - \cfrac{1^6}{|3 \times 39} - \cfrac{2^6}{|5 \times 107} - \cdots$$

$$- \cfrac{n^6}{|(2n+1)(17n^2 + 17n + 5)} - \cdots \qquad \text{(Ex. 9.24)}.$$

In [174] a (complicated) continued fraction for

$$\sum_{n=1}^{\infty} \frac{1}{(a+n)^4}$$

is given, which reduces at $a = 0$ to

$$\text{(A.101)} \quad \zeta(4) = \frac{\pi^4}{90} = \cfrac{13}{|P(0)} + \cfrac{1^7 \times 2 \times 3 \times 4}{|P(1)} + \cfrac{2^7 \times 5 \times 6 \times 7}{|P(2)} + \cdots$$

$$+ \cfrac{n^7 (3n-1)(3n)(3n+1)}{|P(n)} + \cdots \, ,$$

where $P(n) = 3(2n+1)(3n^2 + 3n + 1)(15n^2 + 15n + 4)$. No irregular continued fractions with regular patterns are known for $\zeta(k)$ when $k \geq 5$.

References

[1] B. ADAMCZEWSKI and Y. BUGEAUD, On the Maillet–Baker continued fractions, *J. Reine Angew. Math.* **606** (2007), 105–121.

[2] W. W. ADAMS and J. L. DAVISON, A remarkable class of continued fractions, *Proc. Amer. Math. Soc.* **65** (1977), 194–198.

[3] W. W. ADAMS and M. J. RAZAR, Multiples of points on elliptic curves and continued fractions, *Proc. London Math. Soc.* **41** (1980), 481–498.

[4] G. ALMKVIST and W. ZUDILIN, Differential equations, mirror maps and zeta values, in: *Mirror symmetry V*, AMS/IP Stud. Adv. Math. 38 (Amer. Math. Soc., Providence, RI, 2006), pp. 481–515.

[5] G. ANDREWS and B. C. BERNDT, *Ramanujan's lost notebook*, Parts I, II, II, IV (Springer, New York, 2005, 2009, 2012, 2013).

[6] G. E. ANDREWS, B. C. BERNDT, L. JACOBSEN and R. L. LAMPHERE, *The continued fractions found in the unorganized portions of Ramanujan's notebooks*, Mem. Amer. Math. Soc. 99 (Amer. Math. Soc., Providence, RI, 1992), no. 477.

[7] F. APÉRY, Roger Apéry, 1916–1994: a radical mathematician, *Math. Intelligencer* **18** (1996), no. 2, 54–61.

[8] R. APÉRY, Irrationalité de $\zeta(2)$ et $\zeta(3)$, *Journées arithmétiques de Luminy* (20–24 June 1978), Astérisque 61 (Soc. Math. France, Paris, 1979), 11–13.

[9] D. H. BAILEY, J. M. BORWEIN and R. H. CRANDALL, On the Khintchine constant, *Math. Comp.* **66** (1997), 417–431.

[10] A. BAKER, *A concise introduction to the theory of numbers* (Cambridge University Press, Cambridge, 1984).

[11] J. BARÁT and P. P. VARJÚ, Partitioning the positive integers to seven Beatty sequences, *Indag. Math. (NS)* **14** (2003), 149–161.

[12] A. F. BEARDON and I. SHORT, The Seidel, Stern, Stolz and Van Vleck theorems on continued fractions, *Bull. London Math. Soc.* **42** (2010), 457–466.

[13] B. C. BERNDT, *Ramanujan's notebooks*, Parts I, II, III, IV, V (Springer-Verlag, New York, 1985, 1989, 1991, 1994, 1998).

[14] T. G. BERRY, On periodicity of continued fractions in hyperelliptic function fields, *Arch. Math. (Basel)* **55** (1990), 259–266.

[15] M. R. BEST and H. J. J. TE RIELE, On a conjecture of Erdős concerning sums of powers of integers, Report NW 23/76 (Mathematisch Centrum Amsterdam, 1976).

[16] F. BEUKERS, A note on the irrationality of $\zeta(2)$ and $\zeta(3)$, *Bull. London Math. Soc.* **11** (1979), 268–272.

[17] P. E. BÖHMER, Über die Transzendenz gewisser dyadischer Brüche, *Math. Ann.* **96** (1927), 367–377; Erratum, *Math. Ann.* **96** (1927), 735.

[18] E. BOMBIERI and A. J. VAN DER POORTEN, Continued fractions of algebraic numbers, in: *Computational algebra and number theory*, Sydney, 1992, Math. Appl. 325 (Kluwer, Dordrecht, 1995), pp. 137–152.

[19] D. BORWEIN, J. BORWEIN, R. CRANDALL and R. MAYER, On the dynamics of certain recurrence relations, *Ramanujan J.* **13** (2007), 63–101.

[20] D. BORWEIN, J. M. BORWEIN and B. SIMS, On the solution of linear mean recurrences, *Amer. Math. Monthly* (2014), in press.

[21] J. BORWEIN and D. BAILEY, *Mathematics by experiment. Plausible reasoning in the 21st century*, 2nd edition (A. K. Peters, Wellesley, MA, 2008).

[22] J. M. BORWEIN, D. BAILEY and R. GIRGENSOHN, *Experimentation in mathematics: computational paths to discovery* (A. K. Peters, Natick, MA, 2004).

[23] J. BORWEIN and P. BORWEIN, On the generating function of the integer part: $[n\alpha + \gamma]$, *J. Number Theory* **43** (1993), no. 3, 293–318.

[24] J. BORWEIN, P. BORWEIN and K. DILCHER, Pi, Euler numbers and asymptotic expansions, *Amer. Math. Monthly* **96** (1989), 681–687.

[25] J. M. BORWEIN, K.-K. S. CHOI and W. PIGULLA, Continued fractions of tails of hypergeometric series, *Amer. Math. Monthly* **112** (2005), 493–501.

[26] J. BORWEIN, R. CRANDALL and G. FEE, On the Ramanujan AGM fraction. Part I: the real-parameter case, *Exp. Math.* **13** (2004), 275–286.

[27] J. BORWEIN and R. CRANDALL, On the Ramanujan AGM fraction. Part I: the complex-parameter case, *Exp. Math.* **13** (2004), 287–296.

[28] J. M. BORWEIN and P. BORWEIN, *Pi and the AGM: a study in analytic number theory and computational complexity* (John Wiley, New York, 1987).

[29] J. BORWEIN and R. LUKE, Dynamics of a Ramanujan-type continued fraction with cyclic coefficients, *Ramanujan J.* **16** (2008), 285–304.

[30] J. BORWEIN and R. LUKE, Dynamics of some random continued fractions, *Abstract Appl. Anal.* **5** (2005), 449–468.

[31] J. M. BORWEIN, I. SHPARLINSKI and W. ZUDILIN (eds.), *Number theory and related fields: in memory of Alf van der Poorten*, Springer Proc. Math. and Stat. 43 (Springer-Verlag, New York, 2013).

[32] P. BORWEIN, S. CHOI, B. ROONEY and A. WEIRATHMUELLER, *The Riemann hypothesis: a resource for the afficionado and virtuoso alike*, CMS Books in Math. (Springer-Verlag, New York, 2007).

[33] J. BOURGAIN and A. KONTOROVICH, On Zaremba's conjecture, *CR Math. Acad. Sci. Paris Sér. I Math.* **349** (2011), 493–495.

[34] D. BOWMAN, A new generalization of Davison's theorem, *Fibonacci Quart.* **26** (1988), 40–45.

[35] R. P. BRENT, A. J. VAN DER POORTEN and H. TE RIELE, A comparative study of algorithms for computing continued fractions of algebraic numbers, in: *Algorithmic number theory* (Talence, 1996), Lecture Notes in Computer Sci. 1122 (Springer, Berlin, 1996), pp. 35–47.

[36] E. B. BURGER, *Exploring the number jungle: a journey into Diophantine analysis*, Student Math. Library 8 (Amer. Math. Soc., Providence, RI, 2000).

[37] E. B. Burger and T. Struppeck, On frequency distributions of partial quotients of U-numbers, *Mathematika* **40** (1993), 215–225.

[38] W. Butske, L. M. Jaje and D. R. Mayernik, On the equation $\sum_{p|N} \frac{1}{p} + \frac{1}{N} = 1$, pseudoperfect numbers, and perfectly weighted graphs, *Math. Comp.* **69** (2000), 407–420.

[39] G. Cairns, N. B. Ho and T. Lengyel, The Sprague–Gundy function of the real game Euclid, *Discrete Math.* **311** (2011), 457–462.

[40] D. G. Cantor, Computing in the Jacobian of a hyperelliptic curve, *Math. Comp.* **48** (1987), no. 177, 95–101.

[41] D. G. Cantor, On the analogue of the division polynomials for hyperelliptic curves, *J. für Math. (Crelle)* **447** (1994), 91–145.

[42] D. G. Cantor, P. H. Galyean and H. G. Zimmer, A continued fraction algorithm for real algebraic numbers, *Math. Comp.* **26** (1972), 785–791.

[43] J. W. S. Cassels, *An introduction to Diophantine approximation*, Cambridge Tracts in Math. and Math. Phys. 45 (Cambridge University Press, New York, 1957).

[44] B. M. Char, On Stieltjes' continued fraction for the gamma function, *Math. Comp.* **34** (1980), 547–551.

[45] S. D. Chowla, Some problems of diophantine approximation (I), *Math. Z.* **33** (1931), 544–563.

[46] F. W. Clarke, W. N. Everitt, L. L. Littlejohn and S. J. R. Vorster, H. J. S. Smith and the Fermat two squares theorem, *Amer. Math. Monthly* **106** (1999), 652–665.

[47] H. Cohn, A short proof of the simple continued fraction expansion of e, *Amer. Math. Monthly* **113** (2006), 57–62.

[48] R. M. Corless, G. W. Frank and J. G. Monroe, Chaos and continued fractions, *Phys. D* **46** (1990), 241–253.

[49] T. W. Cusick and M. Mendès France, The Lagrange spectrum of a set, *Acta Arith.* **34** (1979), 287–293.

[50] A. Cuyt, V. B. Petersen, B. Verdonk, H. Waadeland and W. B. Jones, *Handbook of continued fractions for special functions*, with contributions by F. Backeljauw and C. Bonan-Hamada (Springer, New York, 2008).

[51] D. P. Dalzell, On 22/7, *J. London Math. Soc.* **19** (1944), 133–134.

[52] L. V. Danilov, Some classes of transcendental numbers, *Mat. Zametki* **12** (1972), 149–154; English translation, *Math. Notes Acad. Sci. USSR* **12** (1972), 524–527.

[53] H. Davenport, A note on diophantine approximation (II), *Mathematika* **11** (1964), 50–58.

[54] C. S. Davis, A note on rational approximation, *Bull. Austral. Math. Soc.* **20** (1979), no. 3, 407–410.

[55] J. L. Davison, A series and its associated continued fraction, *Proc. Amer. Math. Soc.* **63** (1977), 29–32.

[56] J. L. Davison and J. O. Shallit, Continued fractions for some alternating series, *Monatshefte Math.* **111** (1991), 119–126.

[57] P. Erdős, Advanced problem 4347, *Amer. Math. Monthly* **56** (1949), 343.

[58] S. R. Finch, *Mathematical constants*, Encyclopedia of Math. and its Applications 94 (Cambridge University Press, Cambridge, 2003).

[59] S. FOMIN and A. ZELEVINSKY, The Laurent phenomenon, *Adv. Appl. Math.* **28** (2002), 119–144.

[60] L. R. FORD, Fractions, *Amer. Math. Monthly* **45** (1938), 586–601.

[61] A. S. FRAENKEL, The bracket function and complementary sets of integers, *Adv. Appl. Math.* **28** (2002), 119–144.

[62] A. S. FRAENKEL, Complementing and exactly covering sequences, *J. Combin. Theory Ser. A* **14** (1973), 8–20.

[63] J. S. FRAME, Continued fractions and matrices, *Amer. Math. Monthly* **56** (1949), 98–103.

[64] J. FRANEL, Les suites de Farey et le problème des nombres premiers, *Göttinger Nachrichten* (1924), 198–201.

[65] Y. GALLOT, P. MOREE, and W. ZUDILIN, The Erdős–Moser equation $1^k + 2^k + \cdots + (m - 1)^k = m^k$ revisited using continued fractions, *Math. Comp.* **80** (2011), no. 274, 1221–1237.

[66] A. O. GELFOND, *Calculus of finite differences*, International Monographs on Advanced Math. and Phys. (Hindustan Publishing, Delhi, 1971).

[67] R. L. GRAHAM, Covering the positive integers by disjoint sets of the form $\{[n\alpha + \beta] : n = 1, 2, \ldots\}$, *J. Combin. Theory Ser. A* **15** (1973), 354–358.

[68] R. L. GRAHAM, D. E. KNUTH and O. PATASHNIK, *Concrete mathematics* (Addison-Wesley, Reading, MA, 1990).

[69] D. B. GRÜNBERG and P. MOREE, Sequences of enumerative geometry: congruences and asymptotics. With an appendix by Don Zagier, *Exp. Math.* **17** (2008), 409–426.

[70] R. K. GUY, *Unsolved problems in number theory*, 3rd edition, Problem Books in Math. (Springer, New York, 2004).

[71] D. HANSON, On the product of the primes, *Can. Math. Bull.* **15** (1972), 33–37.

[72] G. H. HARDY and E. M. WRIGHT, *An introduction to the theory of numbers*, 5th edition (Oxford University Press, Oxford, 1989).

[73] H. A. HELFGOTT, *Major arcs for Goldbach's theorem*, Preprint arXiv: 1305.2897v2 [math.NT] (June 2013).

[74] M. HIRSCHHORN, Lord Brouncker's continued fraction for π, *Math. Gazette* **95** (2011), no. 533, 322–326.

[75] A. N. W. HONE, Elliptic curves and quadratic recurrence sequences, *Bull. London Math. Soc.* **37** (2005), 161–171.

[76] A. HURWITZ and N. KRITIKOS, *Lectures on number theory* (Springer-Verlag, Berlin, 1986).

[77] A. E. INGHAM, *The distribution of prime numbers*, Reprint of the 1932 original, with a foreword by R. C. Vaughan, Cambridge Math. Library (Cambridge University Press, Cambridge, 1990).

[78] W. B. JONES and W. J. THRON, *Continued fractions: analytic theory and applications*, Encyclopedia of Math. and its Applications 11 (Addison-Wesley, Reading, MA, 1980).

[79] B. C. KELLNER, Über irreguläre Paare höhere Ordnungen, Diplomarbeit (Math. Inst., Georg-August-Universität zu Göttingen, Germany, 2002); available at http://www.bernoulli.org/~bk/irrpairord.pdf.

[80] A. KHINTCHINE, Metrische Kettenbruchprobleme, *Compositio Math.* **1** (1935), 361–382.

[81] A. KHINTCHINE, Zur metrischen Kettenbruchtheorie, *Compositio Math.* **3** (1936), 276–285.

[82] A. YA. KHINTCHINE, *Continued fractions*, 2nd edition, translated by P. Wynn (P. Noordhoff, Ltd., Groningen, 1963).

[83] D. E. KNUTH, *The art of computer programming*, Vol. II: Seminumerical algorithms (Addison-Wesley, Reading, MA, 1981).

[84] K. KOLDEN, Continued fractions and linear substitutions, *Archiv for Mathematik og Naturvidenskab* **50** (1949), 141–196.

[85] T. KOMATSU, A certain power series and the inhomogeneous continued fraction expansions, *J. Number Theory* **59** (1996), 291–312.

[86] T. KOMATSU, On inhomogeneous Diophantine approximation with some quasi-periodic expressions, *Acta Math. Hungar.* **85** (1999), 311–330.

[87] T. KOMATSU, On inhomogeneous Diophantine approximation and the Borweins' algorithm, *Far East J. Math. Sci.* **12** (2004), 203–224.

[88] T. KOMATSU, A proof of the continued fraction expansion of $e^{2/s}$, *Integers* **7** (2007), no. A30.

[89] C. KRATTENTHALER, Advanced determinant calculus, in: *The Andrews Festschrift* (Maratea, 1998), *Sém. Lothar. Combin.* **42** (1999), Art. B42q, 67 pp.

[90] L. KUIPERS and H. NIEDERREITER, *Uniform distribution of sequences* (Wiley-Interscience, New York, 1974).

[91] R. O. KUZMIN, On a problem of Gauss, *Dokl. Acad. Sci. USSR* (1928), 375–380.

[92] J. C. LAGARIAS and J. SHALLIT, Linear fractional transformations of continued fractions with bounded partial quotients, *J. Théorie Nombres Bordeaux* **9** (1997), 267–279; Corrigendum, *J. Théorie Nombres Bordeaux* **15** (2003), 741–743.

[93] E. LANDAU, Bemerkungen zu der vorstehenden Abhandlung von Herrn Franel, *Göttinger Nachrichten* (1924), 198–206.

[94] S. LANG, *Introduction to Diophantine approximations*, 2nd edition (Springer-Verlag, New York, 1995).

[95] S. LANG and H. TROTTER, Continued fractions for some algebraic numbers, *J. Reine Angew. Math.* **255** (1972), 112–134; Addendum, *J. Reine Angew. Math.* **267** (1974), 219–220.

[96] D. H. LEHMER, Euclid's algorithm for large numbers, *Amer. Math. Monthly* **45** (1938), 227–233.

[97] P. LÉVY, Sur les lois de probabilité dont dépendent les quotients complets et incomplets d'une fraction continue, *Bull. Soc. Math. France* **57** (1929), 178–194.

[98] P. LÉVY, Sur le développement en fraction continue d'un nombre choisi au hasard, *Compositio Math.* **3** (1936), 286–303.

[99] P. LIARDET and P. STAMBUL, Algebraic computations with continued fractions, *J. Number Theory* **73** (1998), 92–121.

[100] F. LINDEMANN, Über die Zalh π, *Math. Ann.* **20** (1882), 213–225.

[101] G. LOCHS, Vergleich der Genauigkeit von Dezimalbruch und Kettenbruch, *Abh. Hamburg Univ. Math. Sem.* **27** (1964), 142–144.

[102] L. LORENTZEN, Convergence and divergence of the Ramanujan AGM fraction, *Ramanujan J.* **16** (2008), 83–95.

[103] L. LORENTZEN and H. WAADELEND, *Continued fractions with applications* (North Holland, 1992).

[104] J. H. Loxton and A. J. van der Poorten, Arithmetic properties of certain functions in several variables. III, *Bull. Austral. Math. Soc.* **16** (1977), 15–47.

[105] S. K. Lucas, Approximations to π derived from integrals with nonnegative integrands, *Amer. Math. Monthly* **116** (2009), 166–172.

[106] K. MacMillan and J. Sondow, Proofs of power sum and binomial coefficient congruences via Pascal's identity, *Amer. Math. Monthly* **118** (2011), 549–551.

[107] K. MacMillan and J. Sondow, Divisibility of power sums and the generalized Erdős–Moser equation, *Elemente Math.* **67** (2012), 182–186.

[108] K. Mahler, Arithmetische Eigenschaften der Lösungen einer Klasse von Funktionalgleichungen, *Math. Ann.* **101** (1929), 342–366; Corrigendum, *Math. Ann.* **103** (1930), 532.

[109] E. Maillet, *Introduction à la théorie des nombres transcendants et des propriétés arithmétiques des fonctions* (Paris, Gauthier-Villars, 1906).

[110] R. Marcovecchio, The Rhin–Viola method for log 2, *Acta Arith.* **139** (2009), 147–184.

[111] J. McLaughlin, Symmetry and specializability in the continued fraction expansions of some infinite products, *J. Number Theory* **127** (2007), 184–219.

[112] N. Möller, On Schönhage's algorithm and subquadratic integer GCD computation, *Math. Comp.* **77** (2008), 589–607.

[113] P. Moree, Diophantine equations of Erdős–Moser type, *Bull. Austral. Math. Soc.* **53** (1996), 281–292.

[114] P. Moree, A top hat for Moser's four mathemagical rabbits, *Amer. Math. Monthly* **118** (2011), 364–370.

[115] P. Moree, H. te Riele and J. Urbanowicz, Divisibility properties of integers x, k satisfying $1^k + \cdots + (x-1)^k = x^k$, *Math. Comp.* **63** (1994), 799–815.

[116] L. Moser, On the diophantine equation $1^n + 2^n + 3^n + \cdots + (m-1)^n = m^n$, *Scripta Math.* **19** (1953), 84–88.

[117] H. Niederreiter, Dyadic fractions with small partial quotients, *Monatshefte Math.* **101** (1986), 309–315.

[118] Ku. Nishioka, *Mahler functions and transcendence*, Lecture Notes in Math. 1631 (Springer-Verlag, Berlin, 1996).

[119] Ku. Nishioka, I. Shiokawa and J. Tamura, Arithmetical properties of a certain power series, *J. Number Theory* **42** (1992), 61–87.

[120] I. Niven, A simple proof that π is irrational, *Bull. Amer. Math. Soc.* **53** (1947), 509.

[121] I. Niven, *Irrational numbers*, Carus Math. Monographs 11, Math. Assoc. Amer. (John Wiley, New York, NY, 1956).

[122] K. O'Bryant, A generating function technique for Beatty sequences and other step sequences, *J. Number Theory* **94** (2002), 299–319.

[123] F. W. J. Olver, D. W. Lozier, R. F. Boisvert and C. W. Clark (eds.), *NIST handbook of mathematical functions* (Cambridge University Press, New York, 2010).

[124] O. Perron, Über die Approximation irrationaler Zahlen durch rationale, *Sitz. Heidelberg. Akad. Wiss.* **12A** (1921), 3–17.

[125] O. Perron, *Die Lehre von den Kettenbrüchen*, 3rd edition, Bd. I: Elementare Kettenbrüche (B. G. Teubner, Stuttgart, 1954); Bd. II: Analytisch-funktionentheoretische Kettenbrüche (B. G. Teubner, Stuttgart, 1957).

[126] M. Petkovšek, H. S. Wilf and D. Zeilberger, $A = B$ (A. K. Peters, Wellesley, MA, 1996).

[127] A. van der Poorten, A proof that Euler missed... Apéry's proof of the irrationality of $\zeta(3)$, *Math. Intelligencer* **1** (1978/79), 195–203.

[128] A. van der Poorten, Formal power series and their continued fraction expansion, in: *Algorithmic number theory*, Lecture Notes in Computer Sci. 1423 (Springer-Verlag, Berlin, 1998), pp. 358–371.

[129] A. van der Poorten, Quadratic irrational integers with partly prescribed continued fraction expansion, *Publ. Math. Debrecen* **65** (2004), 481–496.

[130] A. van der Poorten, Specialisation and reduction of continued fraction expansions of formal power series, *Ramanujan J.* **9** (2005), 83–91.

[131] A. van der Poorten, Elliptic curves and continued fractions, *J. Integer Sequences* **8** (2005), paper 05.2.5, 19 pp.

[132] A. van der Poorten, Curves of genus 2, continued fractions, and Somos sequences, *J. Integer Sequences* **8** (2005), paper 05.3.4, 9 pp.

[133] A. van der Poorten, Hyperelliptic curves, continued fractions, and Somos sequences, in: *Dynamics and stochastics*, IMS Lecture Notes Monogr. Ser. 48 (Inst. Math. Statist., Beachwood, OH, 2006), pp. 212–224.

[134] A. van der Poorten and J. Shallit, Folded continued fractions, *J. Number Theory* **40** (1992), 237–250.

[135] A. van der Poorten and J. Shallit, A specialised continued fraction, *Can. J. Math.* **45** (1993), 1067–1079.

[136] A. J. van der Poorten and C. S. Swart, Recurrence relations for elliptic sequences: every Somos 4 is a Somos k, *Bull. London Math. Soc.* **38** (2006), 546–554.

[137] M. Prévost, A new proof of the irrationality of $\zeta(2)$ and $\zeta(3)$ using Padé approximants, *J. Comput. Appl. Math.* **67** (1996), 219–235.

[138] K. Rajkumar, A simplification of Apéry's proof of the irrationality of $\zeta(3)$, Preprint arXiv: 1212.5881 [math.NT] (2012).

[139] G. N. Raney, On continued fractions and finite automata, *Math. Ann.* **206** (1973), 265–283.

[140] G. Rhin and C. Viola, On a permutation group related to $\zeta(2)$, *Acta Arith.* **77** (1996), 23–56.

[141] R. D. Richtmyer, M. Devaney and N. Metropolis, Continued fraction expansions of algebraic numbers, *Numer. Math.* **4** (1962), 68–84.

[142] T. Rivoal and W. Zudilin, Diophantine properties of numbers related to Catalan's constant, *Math. Ann.* **326**:4 (2003), 705–721.

[143] J. Roberts, *Elementary number theory: a problem oriented approach* (MIT Press, 1978).

[144] A. M. Rockett and P. Szüsz, *Continued fractions* (World Scientific, Singapore, 1992).

[145] V. Kh. Salikhov, On the irrationality measure of π, *Usp. Mat. Nauk.* **63** (2008), no. 3, 163–164; English translation, *Russian Math. Surveys* **63** (2008), 570–572.

[146] A. Schinzel, On some problems of the arithmetical theory of continued fractions, *Acta Arith.* **6** (1961), 393–413.

[147] A. Schinzel, On some problems of the arithmetical theory of continued fractions II, *Acta Arith.* **7** (1962), 287–298.

[148] W. M. SCHMIDT, On badly approximable numbers, *Mathematika* **12** (1965), 10–20.

[149] W. M. SCHMIDT, *Diophantine approximation*, Lecture Notes in Math. 785 (Springer-Verlag, Berlin, 1980).

[150] A. SCHÖNHAGE, Schnelle Berechnung von Kettenbruchentwicklungen, *Acta Informatica* **1** (1971), 139–144.

[151] J. SHALLIT, Real numbers with bounded partial quotients: a survey, *L'Enseignement Math.* **38** (1992), 151–187.

[152] R. SHIPSEY, Elliptic divisibility sequences, Ph.D. thesis (Goldsmiths College, University of London, 2000).

[153] P. SHIU, Computation of continued fractions without input values, *Math. Comp.* **64** (1995), no. 211, 1307–1317.

[154] P. SHIU, A function from Diophantine approximations, *Publ. Inst. Math. (Beograd)* **65** (1999), 52–62.

[155] TH. SKOLEM, Über einige Eigenschaften der Zahlenmengen [$\alpha n + \beta$] bei irrationalem α mit einleitenden Bemerkungen über einige kombinatorische Probleme, *Norske Vid. Selsk. Forh. (Trondheim)* **30** (1957), 118–125.

[156] N. J. A. SLOANE, *The on-line encyclopedia of integer sequences*, published electronically at http://oeis.org/ (2013).

[157] K. R. STROMBERG, *An introduction to classical real analysis* (Wadsworth, 1981).

[158] C. SWART, Elliptic curves and related sequences, Ph.D. thesis (Royal Holloway College, University of London, 2003).

[159] B. G. TASOEV, On rational approximations of some numbers, *Math. Notes* **67** (2000), no. 5-6, 786–791.

[160] R. TIJDEMAN, Exact covers of balanced sequences and Fraenkel's conjecture, in: *Algebraic number theory and Diophantine analysis*, Graz, 1998 (de Gruyter, Berlin, 2000), pp. 467–483.

[161] R. TIJDEMAN, Fraenkel's conjecture for six sequences, *Discrete Math.* **222** (2000), 223–234.

[162] A. J. H. VINCENT, Sur la résolution des équations numériques, *J. Math. Pures Appl.* **1** (1836), 341–372.

[163] J. VUILLEMIN, Exact real computer arithmetic with continued fractions, INRIA Report 760 (INRIA, Le Chesnay, France, 1987).

[164] H. S. WALL, *Analytic theory of continued fractions* (Chelsea Publishing, New York, 1948).

[165] M. WARD, Memoir on elliptic divisibility sequences, *Amer. J. Math.* **70** (1948), 31–74.

[166] E. T. WHITTAKER and G. N. WATSON, *A course of modern analysis*, 4th edition (Cambridge University Press, 1927).

[167] A. J. YEE, γ-cruncher – a multi-threaded pi-program, available at http://www.numberworld.org/.

[168] D. B. ZAGIER, *Zetafunktionen und quadratische Körper* (Springer-Verlag, New York–Berlin, 1981).

[169] D. B. ZAGIER, Problems posed at the St Andrews Colloquium (1996), Solutions, 5th day; http://www-groups.dcs.st-and.ac.uk/~john/Zagier/Problems.html.

[170] D. ZAGIER, Vassiliev invariants and a strange identity related to the Dedekind eta-function, *Topology* **40** (2001), 945–960.

[171] D. ZAGIER, Integral solutions of Apéry-like recurrence equations, in: *Groups and symmetries*, CRM Proc. Lecture Notes 47 (Amer. Math. Soc., Providence, RI, 2009), pp. 349–366.

[172] S. K. ZAREMBA, La méthode des 'bons treillis' pour le calcul des intégrales multiples, in: *Applications of number theory to numerical analysis*, Proc. Sympos., Université de Montréal, 1971 (Academic Press, New York, 1972), pp. 39–119.

[173] Y. ZHANG, Bounded gaps between primes, *Ann. Math.* (2013), in press; http://annals.math.princeton.edu/articles/7954.

[174] W. ZUDILIN, Well-poised generation of Apéry-like recursions, *J. Comput. Appl. Math.* **178** (2005), 513–521.

[175] W. ZUDILIN, Apéry's theorem. Thirty years after, *Intern. J. Math. Computer Sci.* **4** (2009), 9–19; An elementary proof of Apéry's theorem, Preprint arXiv: math.NT/0202159 (2002).

[176] W. ZUDILIN, On the irrationality measure of π^2, *Usp. Mat. Nauk.* **68** (2013), no. 6, 171–172; English translation, *Russian Math. Surveys* **68** (2013), 1133–1135; Two hypergeometric tales and a new irrationality measure of $\zeta(2)$, Preprint arXiv: 1310.1526 [math.NT] (2013).

Index